The Medium is the Message, Explained: A Guide to Understanding Media Effects

Meenu

Table of Contents

Chapter 1: Introduction

What is a medium? For starters, it is a concept that takes many forms. T-shirts, soothsayers, sculptures, and screens are all valid examples. Blood is a medium that carries oxygen in our bodies, a pipe is the medium by which water is brought into our homes, and words are the medium by which we express our thoughts and ideas. In all these cases, the definition relates to something that lies "in-between" or in the middle of larger processes. Being in the middle, a medium is a means for making connections. That is, media are the facilitators, intermediaries, middlemen, or portals *thorough* which actions and/or communications occur.

Importantly, they are not the actions and/or communications they carry. The television is not the gameshow it displays, nor is the Internet the music video it streams. It is this distinction that makes a discussion of media effects possible. Sometimes, media effects are meant to mean the outcomes related to being exposed to the content (i.e. messages/information) carried by media to their users. One might talk about the effect of violent video games on youth behavior or how a news organization's reporting influences popular opinion of a public official. These issues are important, but they don't address the influence that media have as intermediaries on the individuals who use them.

Annie Lang (2013) argued for a broader understanding of media as agents that fundamentally change society and culture. Rather than looking at individual instances of content consumption, she suggests the field of mass communication needs to shift its focus to understanding how access to various media/portals/channels are psychologically relevant to individuals in the modern era. In other words, how do media (i.e., portals through which we access and send information) affect our psychological perceptions of ourselves and others?

It is with this broader understanding of media effects in mind that the following study investigates how individuals' use of "smartphones" (mobile devices that combine telephony and computing capabilities) may be related to perceptions of privacy, presence, and publicity. Smartphones are an important medium for study because despite being a relatively new technology, they have quickly become one of the primary (if not the primary) medium for content consumption and dissemination. In 2009, roughly one-third of adults in the United States owned a smartphone. Today, that number has risen to more than 80% and is over 90% for millennials (Pew Research Center, 2019). Unlike previous mass media, the modern smartphone comes equipped with a wide range of sensor technologies (e.g., GPS, cameras, microphones, thermometers, to name a few). These sensors give users both an unprecedented range of potential functions and potentials for being surveilled (Chester, 2012). The mobility of smartphones is another difference that sets it apart as a medium. Before smartphones, individuals needed to locate themselves near a medium to consume content. With smartphones, they have access to an essentially infinite universe of content regardless of where they are. As a result, two related industries have proliferated with the rise of smartphones. First, a big data industry which functions to collect and sell information about users to marketers and advertisers. Second, a mobile application industry whose business models are based on attracting and retaining user's attention. That is, smartphones have simultaneously generated unprecedented potentials for gathering information about preferences, tastes, and daily lifestyles and the ability for individuals to consume and engage with media content at virtually every moment of their lives. Given these developments, it is imperative to understand how this recent and vast increase in connectivity is impacting individuals and society. In order to contribute to this overall understanding, the following chapters contain relevant prior research and a novel research project designed to investigate how millennials are managing privacy, presence, and publicity in the face of digital surveillance and "always on" connectivity?

It begins with an overview of theories related to media effects and how they may be applied and understood in relation to the internet and/or cyberspace and proceeds to an examination of empirical studies related to the effects on privacy, publicity and attention happening because of the growing digital connectivity made possible by modern smartphones. Through the examination, smartphones come to be understood as giving individuals the ability (and likely the requirement) to manage both physical and digital presences as a part of their everyday experience. As a result, individuals engage in a negotiation between public and private identity which leads to attendant concerns about performance, autonomy, and safety. Importantly, it also requires them to allocate their finite resources of time and attention between the digital and physical worlds in which they exist. What results is a complicated notion of identity that is continually being negotiated through one's interactions with individuals and environments that increasingly blur the lines of intended and unintended communication behaviors as well as notions of the public and private individual (Auxier et al., 2019; Boyles et al., 2012). Individuals are aware smartphones communicate information about them without their knowledge and at times may collect personal information they would rather keep private. However, they also report regularly using their smartphones for activities perceived to jeopardize the security of their personal information in return for gains in social capital and self-esteem.

A repeated theme in this research is the tension between user concerns about how their private information and attention resources are allocated and smartphone applications that compete for user engagement and are always on hand for users' inquiries and general entertainment. This builds upon previous research that shows users perceive substantial gains in social connectivity, information access and productivity (Acquisti et al., 2015; Keith et al., 2014; M. Mazmanian & Erickson, 2014; Xu et al., 2009). However, these gains come with costs to personal privacy, their ability to focus attention, and relationships in face-to-face settings (Brandimarte et al., 2013a; Rodriguez et al., 2018; Rosenberger, 2020; Sobhani & Farooq, 2018; Throuvala et al., 2020).

A theme throughout this study is a loss of privacy both in the sense of controlling one's personal information and in the ability to be truly alone because of nearly ubiquitous smartphone use. The connectivity made possible by smartphones offers unprecedented access to others and information at virtually all times and in all places. This gives users great benefits in terms of convenience but carries with it risk to their personal autonomy. There is endless access to information. Entire libraries of books, newspapers, movies and songs are available at the touch of a button. Similarly, one may contact endless numbers of people. Family, friends, teachers, and service providers of all sorts may be reached with a call, text, post and/or video chat at all times of the day regardless of one's location. Importantly, this ability to contact others and connect with information also means individuals can be contacted by others and organizations. This two-way street becomes a serious drain on users' finite time and attention resources. Furthermore, the sensors common to most modern smartphones combined with internet tracking technologies (e.g. "cookies"), has given rise to a big data industry capable of profiling and tracking users in ways that are often beyond users ability to fully comprehend and manage (Reidenberg et al., 2015).

Indeed, the argument is made that many companies either purposely or negligently hide their information sharing practices behind vague and convoluted permission requests and privacy policies. Even well informed and privacy conscious users have been shown to be incapable of deciphering many of these policies and many more users have come to ignore them altogether. Given the multitude of demands on most users' time, this lack of capability is understandable, but it only serves to compound the issue as smartphone applications will continue to battle for users' attention and leave users with even less time to sufficiently evaluate and understand the policies associated with the applications they use daily.

A basic question to start with is how did we get to a point where the most ubiquitously used tool in the modern environment may ultimately be undermining our abilities to perform tasks well and develop meaningful, deep relationships? In the chapters that follow, insights into this question and several proposed tools and teachings for harnessing the good and minimizing the negative impacts of smartphone use will be presented. In total, the results generate more questions than they answer, however the generation of new questions is a valuable resource for deeper understandings in the future.

This dissertation is organized as follows. In the next chapter theories of media effects are presented as a means for creating predictions about the effects of smartphones. Then, in Chapter 3, empirical studies related to the impact of media on privacy, publicity, and the ability to manage one's attention (i.e., presence) are reviewed to identify gaps in the literate that the present study can address. After identifying the study's primary research question at the conclusion of Chapter 3, Chapter 4 presents a survey methodology that was used to test the study's hypotheses and generate variables for inferential statistics. The presentation of the study's results is divided into two chapter. In Chapter 5, descriptive statistics of survey response frequencies are presented to give a general overview of participants' beliefs and smartphone behaviors. Then, in Chapter 6, inferential statistics and regression models generated by the responses are presented that can be used for generalizing about the sample's populations. Chapter 7 puts the study's results in the context of previous literature and speculates about a theoretical model of smartphone use that may be useful for testing smartphone effects in future studies. Finally, Chapter 8 reviews key information and themes developed throughout the study and concludes the dissertation.

Chapter 2: Media Effects, Privacy and Cyberspace Theory

Before investigating the effects of modern media like smartphones, it is useful to explore media theories that make predictions about how different forms of media may affect individuals and society. The following chapter moves chronologically from some of the earliest media effects theories focused on print and broadcast media to modern theories of cyberspace and embedded computing environments. In doing so, it presents historical theories (principally that of Marshall McLuhan) that can be evaluated for their relevancy to the modern environment. The chapter also provides useful terms and concepts for investigating media that have transformed historical understandings of mass communication as an information exchange between a producer and a mass audience to a model where all nodes in a network may be correctly characterized as producers and audience members. The terms and concepts discussed in this theoretical literature review are ultimately utilized in the development of the study's research questions and variables. Many of the theoretical terms, concepts and predictions presented below are reflected upon frequently in the discussion and conclusion chapters.

Media Effects Theory and Background

The study of media effects is relatively new. Although timelines vary, the work of Katz and Lazarsfeld (1955) is often cited as the first formalized theory of media effects produced in the US (Pietilä, 2005). Like much of the effects research that would follow, their work was interested in understanding the extent to which mediated messages (i.e. content) influenced the ideas and opinions of audiences.

Understanding the effects of content consumption is important. It has enhanced our understanding of how propaganda, advertising messages, and gender/class/sexuality/race representations portrayed in entertainment products impact our lives. Furthermore, it has drawn attention to important political and social inequalities that exist within the media industries. However, studies focusing on media content often omit or ignore discussions of the content delivery devices that make consumption possible (Caporino et al., 2020; Chen et al., 2020; Krongard & Tsay-Vogel, 2020). In an era characterized by the rapid introduction and adoption of new media devices (phones, tablets, "wearables," virtual reality displays), it is imperative scholars seek to understand how different media (e.g. books, newspapers, radios, televisions, computers, smartphones, tablets, etc.) impact our ways of existing in and interacting with the world.

Marshall McLuhan was among the first academic researchers to make "medium effects" the center of his work. Famously asserting, "the medium is the message/massage," he argued, "the things on which words were written down count more than the words themselves" (McLuhan, 1964). While debated to this day, his work is important for drawing attention to the channels or means of communication rather than the content delivered through them.

McLuhan's theory for understanding media began with envisioning every medium as an extension of at least one of the human senses. For example, he argued the radio extended one's sense of hearing. With the aid of a radio, audiences could listen in on content being broadcast from miles away. Similarly, a picture extends our sense of sight. A picture allows us to see objects and scenery in places physically too far away from our location to be viewed using our eyesight alone. In this fashion, McLuhan argued media extend (and thereby focus and hone) our various senses. Building on this line of thinking, his theory of media effects sought to explain how extending our senses through different media impacted the ways we thought about ourselves and the world around us (McLuhan, 1964).

Envisioning tribal/primal societies (or societies built largely upon interpersonal relationships within small geographic spaces) as the baseline for human experience, McLuhan observed that "new (media) set up new equilibrium(s) among all of the senses and faculties leading to… new attitudes and preferences in many areas" (McLuhan, 1964, p. 125). The specific attitudes and preferences created depended on the "the new ratios, not only among our private senses, but among (media) themselves" established each time a new medium was adopted by society. McLuhan argued every medium favored and generated one of two diametrically opposed effects: explosion and implosion.

Explosive Media Effects

Any medium that has the effect of moving the exchange of information away from a tribal or collective center, McLuhan labeled "explosive." Perhaps his clearest and most compelling example of an explosive medium was the written/printed phonetic alphabet. The phonetic alphabet is a "more efficient means for storing and expediting information" (McLuhan, 1964, p. 158) than the spoken word or other early forms of written communication like logographs and hieroglyphs. It was the efficiency (i.e. ability to be learned/applied quickly) of the phonetic alphabet that led to widespread literacy.

Combined with paper and the printing press, mass literacy allowed information to be communicated across expansive physical geographies. The printed phonetic word caused individuals to move away from tribal centers ("explode") because it allowed for information to be stored, transported, and consumed by individuals separated by large physical distances. In other words, it allowed information to travel efficiently without loss of fidelity.

In addition to increasing the distances across which information could travel, the written word affected the perspectives of its users. McLuhan theorized the dominance of the written/printed word for nearly two-thousand years led to many of our current practices, values, and beliefs. In his words, "the separation of the individual from the group in space (privacy), and in thought ("point of view"), and in work (specialism), has had the cultural and technological support of literacy" which in turn created a society characterized by "fragmented industrial and political institutions" (McLuhan, 1964, p. 107). To better understand this thinking, it is useful to consider how written/printed media alter the manner in which information is consumed and presented.

As an individual activity, reading requires spaces in which individuals can be "alone with their thoughts." McLuhan believed the need to mentally separate from our environments while reading generated our earliest notions of "private" property. Written/printed media deliver a singular point of view (the author's), and as widespread literacy generated more and more consumption of these singular viewpoints, so too grew a sense of individualism within readers and authors alike. Also, the invention of libraries, offices, and classrooms were the result of individuals needing spaces for contemplation in communities where the printed phonetic word was/is the primary means of information exchange (McLuhan, 1964).

Print media also organize information by subject or topic. This allows information to specialize, which McLuhan noted is what allows individuals to specialize or focus their knowledge on specific subjects and proficiencies. All modern books with the explicit purpose of teaching (i.e. textbooks) are specialized volumes concerning individual topics like medicine, engineering, geography, psychology, music, painting, literature, etc. Winnowing information into specializations is what underpins the division of labor in developed societies. That is, professionals are the products of specialized training that relies on the separation of concepts into categories that create *individual* fields of study (McLuhan, 1964).

McLuhan further attributed notions of homogeneity and permanence to literacy. Once published, a printed work's text remains unchanged and can be copied in perpetuity. As information and ideas were understood to originate with individuals who could share *their* thoughts with others, it followed that there needed to be a means for protecting the ownership and control of point-of-view (i.e., copyright). Lasting individual ownership of information further expanded definitions of private property and individuals' rights to control how *their* information was consumed and utilized by others in the community (McLuhan, 1964).

To summarize, McLuhan's theory attributed the modern notions of individualism, privacy, specialization, and continuity to the widespread literacy made possible by the "explosive" printed word. The printed word "exploded" the collective, plural and present minded "tribal" societies that adopted it and transformed them into the individual, private and specialized societies we see today. In the wake of this transformation, were the values and preferences for privacy (information control), ownership (i.e. private property) and individualism (i.e. freedom/autonomy) displayed by many Western societies today.

Implosive Media Effects

Although it was the increased efficiency by which print media facilitated consumption that led McLuhan to argue its "explosive" effects, he ultimately believed the efficiency of electronic media would "implode" societies back to their tribal origins (McLuhan, 1964). He argued the speed and volume of information exchanged through electronic media (principally radio and television at the time of his theorizing) was so fast that individuals would sense being in communion with one another regardless of physical separation in space or time. That is, he predicted electronic media generating "implosive" or "retribalizing" effects that would work to counter the effects of print. Using his conception of media as "extensions of the senses," he thought electronic media (especially television) had the effect of extending one's entire central nervous system rather than a specific sense like hearing or touch. "When information moves at the speed of signals in the central nervous system, man is confronted with the obsolescence of all earlier forms of acceleration… What emerges is a total field of inclusive awareness" (McLuhan, 1964, p.104). McLuhan predicted the "inclusive awareness" generated by the almost real time exchange of information between individuals across the globe with electronic media would have the effect of imploding or collapsing societies back to their tribal origins. "In the electronic age, instant speeds abolish time and space, and return man to an integral and primitive awareness" (McLuhan, 1964, p. 152).

"Experiencing" information through electronic media is the antithesis of consuming information through "explosive" media like the printed word. The "participatory" nature of electronic media lessened the need for individuals to consume information in isolation from one another. Families and friends could gather around their radios and television sets. Although the phonetic alphabet promoted widespread literacy, radio and television content required no specialized/formal training to consume. Rather than reinforcing ideas of homogeneity, privacy, permanence, and ownership/control, McLuhan argued electronic media supported cultural pluralism, uniqueness, and discontinuity. He thought electronic media would ultimately generate a communal consciousness, "in the electronic age, we wear all mankind as our skin... The subliminal life, private and social, has been hoicked (sic) up into full view, with the result that we have 'social consciousness'" (McLuhan, 1964, p. 47).

McLuhan theorized the "social consciousness" resulting from communications through electronic media would have the effect of creating a "global village." Instead of a world dominated by societal or cultural centers that extend the margins or boarders of their influence/control through printed media, McLuhan foresaw electronic media ushering in a time of multiple societal/cultural centers colliding, overlapping, competing, and collapsing with one another. "Electronic speeds create centers everywhere. Margins cease to exist on this planet" (McLuhan, 1964, p. 91). That is, rather than societies underpinned by core values and practices that individuals internalize to become members of the culture, McLuhan foresaw the fragmenting of society back into multiple heterogeneous small cultural tribes maintained through the diversity of representation and information sharing experienced through electronic media.

McLuhan's vision may (and has) been debated (Ferguson, 1991; Fishman, 2006; Rosenthal, 1968), but his theory is important to modern study because it draws attention to the potential for media to alter individuals' understandings and perceptions of their access to information and others in their "immediate" environment. Of central interest is how individuals are managing their accessibility to information and others in an era where mobile electronic communication devices (i.e., electronic media that are generally "always on" and highly portable) have become pervasive. How are mobile communication devices affecting individuals' sense of presence or attention to others, how is this sense of presence impacting views of privacy/visibility to others and inversely how may it be impacting or generating perceptions of publicity or public identity? Alternatively, media technologies not only change how information is consumed but also how information producers and consumers are connected in time and space. This distinction was central to the work of Walter Ong.

The Mediated Audience

The movement from cultures dominated by aural traditions to those dominated by print changed not only the way information was received or consumed by audiences, but also the thought processes and strategies of those who desired to send or produce information. In oral communication, information senders and receivers occupy synchronous presence in time and space. The separation of sender and receiver created by the written word (and all media technologies) fundamentally alters how information is communicated by removing the immediate feedback of audience members (i.e., receivers). The consequences of this separation were at the heart of Walter Ong's scholarship (Ong, 1975, 1979).

Ong pointed out that direct communication through the printed word is impossible and requires authors to imagine their audience members. Somewhat ironically, literacy had the effect of making both authors and audience members desire solitude. As Ong demonstrates, "I want to write a book which will be read by hundreds of thousands of people. So, please, everyone leave the room. I have to be alone to communicate" (Ong, 1979, p. 3). By separating the author from her audience, Ong argues authors are forced to imagine topics and events that will be of interest or significance to an audience of unknown and unseen others. The result of which is that authors frame contexts through which their texts are intended be read and audience members are called on to play the roles or adopt the lenses authors provide. The degree to which authors and audiences successfully frame and interpret such contexts generally drives the level of perceived success of any given work (i.e., generally evaluated by sales and/or number of readers). That is, the success of mediated communication can be thought of as a reflection of the authors ability to correctly identify topics of interest to a wide range of individuals and compelling frame these topics so that audience members feel that the topic has been fully, but not cumbersomely presented to them. This is no easy task, as again, the author is deprived the feedback made possible by elemental human communication via face-to-face exchanges. Yet, what happens when information senders or authors are asked not only to imagine their potential consumers, but contend with potential surveillance entities or eavesdroppers? The rise of smartphones has created questions concerning accessibility to others in one's personal and professional networks (i.e. those in their desired audience) and their accessibility to unknown third parties and to technology itself (i.e. advertisers, application manufactures, government institutions and algorithms). In this context, the work of Joshua Meyrowitz (1985), who investigated how perceptions or senses of "place" were altered because of electronic media is illuminating.

Performance, Place/Publicity and Electronic Media

Meyrowitz (1985) combined McLuhan's theorizing with the work of Erving Goffman (1959) to highlight how meanings of "place" evolved as a result of electronic media. Among his observations were electronic media destabilizing perceptions of "mainstream" or acceptable performances of masculinity/femininity, childhood/adulthood, and public/private behaviors. Individuals began to question their "place" in society because of historically hidden or restricted information maintained by the specialized categorization and containment of information in print media becoming accessible to mass audiences through electronic media. Children were inadvertently exposed to programming designed for adults, males may watch soap operas targeted to female audiences, motion pictures invited women into men's offices and locker rooms and politicians were held accountable for every word and gesture they uttered because of video recording technologies and the television broadcasts. The influx of information created by electronic media required individuals to adapt their behaviors to reflect the new "collective knowledge" of electronic communities (e.g., broadcast news content and popular entertainment programs). Meyrowitz believed these new behaviors would lead to new definitions of identity that would challenge those generated during the dominance of print.

Building on Goffman's (1959) belief that all social interactions were better characterized as "performances," Meyrowitz maintained all human behavior could be classified as taking place in "front regions" ("front-stage") or "back regions" ("backstage"). "In front regions, the performers are in the presence of their 'audience' for a particular role, and they play a relatively ideal conception of a social role" (Meyrowitz, 1985, p. 29). For example, a politician giving a speech at a campaign rally will be in a front region while "performing" to her political audience. Consequently, she may wish to communicate a demeanor of calm, commanding, and creditable political servant to potential voters. When she returns home and is no longer in the presence of her political audience, she may perform the role of a loving and nurturing partner. Goffman did not assert one performance was more authentic than another, rather the need to notice that behavior is performance based on a performer's understanding and beliefs about what is expected of her by her audience.

In contrast to front region situations, back region situations are characterized by the absence of a particular audience (i.e., private regions). The politician rehearsing her speech before a rally with campaign advisors could be called back region behavior because her electoral audience is not present. On the other hand, she will be performing front region behaviors to her advisors (perhaps joking about mistakes she makes while rehearsing or questioning the wording of her remarks with a speechwriter). Again, what is important to notice is the change in behavior because of an audience being present or absent from any given performance/behavior.

In the age of electronic media, private spaces or back regions are becoming increasingly difficult to maintain. Electronic media permeate spatial and cognitive boundaries that print media could not. Radio and television waves are broadcast over "public airwaves" in the United States. This terminology is not a coincidence. Broadcast frequencies are "public" precisely because they are always available to all people (provided one has a receiver capable of processing the broadcast signals). Electronic media waves surround us everywhere we go, and we have no way of "removing" them from our homes, offices, or classrooms. The pervasiveness of electronic media signals combined with populations characterized by (nearly) ubiquitous adoption and utilization of mobile media receivers (e.g. smartphones and internet connected home assistants) has created a new dimension of experience in which individuals are present not only in their physical surroundings, but also in/through digital connectivity and physical environments which incorporate digital surveillance and interactive capabilities. The proliferation and evolution of digital communication technologies has enhanced or amplified our abilities to interact with our environment(s) and with information networks while making it almost impossible to "turn off" our mediated relationships with these informational spaces/places.

Cyberspace and Cyburg

The modern media environment is increasingly dominated by participation in a digital realm sometimes described as "cyberspace." Cyberspace is a term for which most individuals will have an intuitive understanding, but is nonetheless nebulous in nature. It has been defined as "having no physicality, no matter, and no Cartesian duality, because there is only the mind, our communication is the only transaction" in cyberspace (Cuff, 2003, p. 44). Although lacking physical or geographic space, individuals often speak about meeting on Skype, seeing someone on Facebook, or finding a friend online. Indeed, the uniform resource locator (URL) or internet address was one of the fundamental elements Tim Berners-Lee created when developing the World Wide Web (Hanson, 2011, p. 352). Since its inception, we have understood the web as a sort of digital space where locations exist independent of physical coordinates to host them. Rather than limiting its existence, the lack of physical boundaries in combination with mobile accessibility has led to many individuals occupying cyberspace while being physically located almost anywhere in the world and for long durations of time (provided they have a signal).

As mobile communication technologies/media have evolved to give access to digital worlds regardless of where one stands in physical or geographic space, a third sort of space has sprung up that Cuff (2003) terms the "cyburg" (p. 44). She explains, "If cyberspace is dematerialized space, the cyburg is spatially embodied computing, or a (physical) environment saturated with computing capability" (p. 44). Today, most American's live in the cyburg more often than not, and mobile computing and digital sensing technologies continue to grow in sophistication. With wearable and virtual simulation technologies being innovated and introduced to the market by companies including Google, Amazon and Apple, it is reasonable to conclude mobile communication media are poised to become even more pervasive in the coming years (Ranger, 2020).

As this occurs, we will increasingly come to exist simultaneously within cyberspace and the cyburg. Alternatively, we will occupy what Cuff calls "enacted environments… (where) not only do the walls have ears, but networks of eyes, brains, and data banks to use for purposeful action" (Cuff, 2003, p. 44). Living in these environments will enhance individuals' agency though greater access to information (i.e. leading to more informed decision making) while decreasing individuals' abilities to control the surveillance of their behaviors. Enacted environments will be full of miniaturized sensory and processing technologies that, while invisible to human perception themselves, will make formally unobservable elements in the environment visible to individuals occupying them.

In this fashion, "the cyburg enhances individuals' agency (because) they can know and act in more powerful ways" (Cuff, 2003, p. 44). The success of social networking sites like Facebook have often been attributed to the gains in social capital (i.e. increases in one's social agency) generated by those who utilize these services (Ellison et al., 2007). However, the increased agency made possible by modern media devices also generates an unprecedented amount of information about their users. In response, "what was once considered private is integrated and exposed in public – our intimacies (e.g. cameras that watch bedrooms and bathrooms on reality TV) and our secrets (e.g. medical, legal, and financial databases linked to a national identity card)" (Cuff, 2003, p. 46).

In pervasive computing environments, Cuff (2003) suggests the continuum model described by Meyrowitz (1985) as moving from private to semi-public to public "might instead be replaced by a nested metaphor in which publicity has infected privacy in every conceivable context, and vice versa" (p. 47). Much like Meyrowitz's argument that TV and radio content were bringing conversations that once took place "behind closed doors" between those of similar ages, genders, races, and social classes to the mass public at large, Cuff suggests pervasive computing environments should be viewed as "dismembering mechanisms of modernity (or) mechanisms that break apart social relations across space and time, and that remove local control of resources, services, information, and even the mechanisms themselves" (p. 47).

The "hyper-connectivity" of pervasive computing environments may level the playing field in terms of access to information, but Cuff cautions that pervasive computing environments are realms of "dispersed displacement." In such environments, "discourse about centers and margins becomes irrelevant. For lovers walking hand in hand while speaking simultaneously by cell phone to their respective spouses, spatial dislocation is crucial and unquestioned. In this they remain secure. But they cannot be certain even about the immediate other: with whom is she speaking? Is she with me, or is she elsewhere? In this context, the other (person) is not just distracted; neither is she absent. Instead, she is both present and absent in a way that was not possible prior to wireless technologies whereby everywhere is connected" (Cuff, 2003, p. 48).

The ability to be both present in and absent from one's environment is not unique to modern times. Daydreaming is presumably as old as human existence. For centuries individuals have been transported from their physical locations into worlds of imagination through novels, newspapers, and periodicals. Drawings, sculptures, and photography allowed individuals to lose themselves in visual imagery for hundreds of years. Rather, it is better to say that digital media have been amplifying or enhancing individuals' sensory engagement with distant and potentially fictional locals and individuals through the sounds transmitted by the radio and telephone, the images of the television, and the multi-media representations found online.

Semantics aside, it has become increasingly difficult to assess if an individual's attention is in one place or another in an era of pervasive electronic media. In part, this difficultly stems from the substantial sensory engagement individuals may have with other individuals and environments far removed from their physical locations in geographic space. Skype conversations and interactions in virtual worlds like The World of Warcraft allow individuals to be "present" in ways not possible before the advent of web.

The connectivity of pervasive computing environments also has the potential to bombard individuals with a nearly constant stream of information, alerts, messages, and offers. The picture of a family at the dinner table with everyone (including the baby) on their smartphone has become a common meme (Neje, 2016), and it is not uncommon to witness sidewalk collisions between individuals engaging with their mobile media devices (not to mention while driving). In sum, the notion of "presence" has been problematized by the proliferation of mobile electronic media devices and the connectivity they create between their users and networks (i.e. social relationships, databases, companies, and/or organizations).

"Presence" in the Cyburg

In response to new media environments, Yu & Shaw (2008) attempted to classify the new notions of presence being created. Recognizing that digital and physical spaces are not independent of each other in the cyburg, Mokhtarian's (2002) relationships between the physical and digital realms are used to illustrate the varying ways these two different, but often overlapping, spaces interact. Primarily, digital space may replace or *substitute* its physical counterpart as when individuals telecommute to work, use E-banking, or teleconference. Alternatively, the digital and physical may *complement* each other as when a mobile social networking platform allows individuals to arrange a spontaneous face-to-face meeting. Thirdly, digital media may *modify* former physical transactions. For example, purchases made online modify the means by which the product is delivered to the consumer. Finally, there may be instances where digital media have no influence (this is becoming increasing rare, and may soon be nonexistent) on physical reality, and Mokhtarian calls these *neutrality* relationships.

With Mokhtarian's relationships as a foundation, Shaw and Yu (2008) devised a means by which individuals' presence in the digital and physical worlds could be represented by geographic information systems. In doing so, they identified the potential for individuals to have four different kinds of presence in pervasive computing environments: 1) Individuals may have "synchronous presence" when they coincidence in both space and time. This is equivalent to notions of face-to-face interactions or being physically present with another individual. 2) Individuals have "asynchronous presence" when they coincidence in space, but not in time. For example, two individuals could physically visit the same historical landmark, but on different days of the week. 3) In the digital realm, individuals have "synchronous tele-presence" when they coincide in time, but not in space." Such presence occurs when two individuals "meet" on Skype. Although hundreds or even thousands of miles may separate their physical locations, they are able to share each other's company in real time through mediated representations of the other. 4) Individuals may share "asynchronous tele-presence" even if they do not coincide in space or time. For example, individuals corresponding through wall posts on one another's Facebook pages have asynchronous tele-presence. Despite the lack of real-time communication, and never sharing the same physical location, such individuals nonetheless foster a sense of "presence" in one another's lives.

Although these categories are helpful for understanding how notions of presence are evolving in response increasingly pervasive computing technologies, they also seem inadequate in capturing the true diversity of potential presences one may be maintaining at any given moment in the modern media environment. Mobile "smart-phones" are constantly emitting information about one's location, conversations, and activities. Many individuals maintain multiple digital representations of themselves on various social networking platforms. Mobile social networking platforms encourage users to "check-in" as they travel through the physical world. Musicians complain about audience members experiencing their concerts though their mobile recording devices, rather than being "present" to the performance. Digital environments are allowing individuals to have "invisible" presences (invisible from other users at least) within them, a reality that has given rise to the concept of digital stalking, and most individuals are always available to receive email, news, social media, and personal correspondence notifications via their mobile devices at any hour of the day or night.

Collectively, the variety of "presences" made possible by modern electronic media technologies have led to a proliferation of academic research concerning privacy and digital identity management. Much like McLuhan predicted, we are struggling to balance our values for privacy and individualism in an era of pervasive monitoring/surveillance technologies that allow us to act and coordinate with others and information about ourselves in ways never possible until now. The degree of control individuals, governments, and business should have over the information generated in pervasive computing environments has consequently become an imperative and highly contested issue across the globe.

This chapter has presented media effects theories that predict and explain how media may change perceptions of privacy, presence, and publicity. These theories are important to the present study because they provide a basis of expectation for how smartphones may influence individuals. The immediacy and constant connectivity of smartphones greatly increases the amount of information users have access to and the amount of information available about them. McLuhan (1964) and Meyrowitz (1985) would predict this influx to alter individuals' expectations for knowledge and information. They would expect individuals to be simultaneously more aware and considerate of others outside their "home" cultures and overwhelmed by the knowledge of so many differing points-of-view. Yet, this chapter also goes a step farther. As the internet has given birth to cyberspace and the hybrid cyburg space, individuals now maintain multiple representations of themselves on social media and other internet platforms. This opens many doors for exploring how identity and social consciousness will evolve in response to new media technologies, but the present study confines its thinking to questions of information control (privacy), identity performance (publicity) and attention management (presence). In the next chapter, previous work in all these arenas will be reviewed to support and provide a foundation for this study's investigation and findings.

Chapter 3: Empirical Studies of Privacy, Presence & Place

Whereas the preceding chapter reviewed theoretical literature related to the relationships between media, privacy, presence, and publicity, this chapter will review historical and modern empirical research in the same vein. It begins by establishing a conceptual understanding of what privacy means in addition to its application in American legal contexts. Privacy is suggested to be a vague or contested concept made more so by smartphone sensors and internet tracking technologies. Modern efforts to enhance or protect smartphone user privacy are discussed in relation to threats created by machine learning and the big data industry. Ultimately, privacy and its inverse publicity are shown to be central to current debates about the risks and benefits of smartphone use. Since most of the privacy protection tools seen in the literature have not been adopted by the public, how individuals are responding to "always on" accessibility and having their private information sold and shared by various third parties is explored by studies presented at the end of the chapter. These studies will highlight aspects of the attention economy and the impact interruptions from smartphone notifications and alerts have been shown to have on task performance and focus. The chapter concludes with research questions generated by the literature presented by this and the preceding chapter.

Privacy: Origins and Evolution

Academic studies of privacy are not new, nor are the exclusive to the field of communication. Klopfer & Rubenstein's (1977) discussion of privacy draws attention to the fact that territoriality is a feature common to virtually all animal species. There is a universal desire of animals to have a space of their own and they will defend this space (sometimes at great personal cost) against intrusions from others. This is also a useful starting point for a definition of privacy "as a regulatory process that serves to selectively control access of external stimulation to one's self or the flow of information to others" (p. 53). The literature today is abounding with definitions that isolate privacy into specific contexts and/or describe the contested nature of the term (Könings et al., 2014; Mulligan et al., 2016; Ziegeldorf et al., 2014), but the simplicity, breadth and primacy of Klopfer & Rubenstein's (1977) definition is useful for a high level understanding of the concept. Recognizing privacy as something desired by all animals, we can hone our thinking about how smartphones are affecting desires for and understandings of privacy within a spectrum. Drawing on the McLuhan discussion above, explosive media do not create desires for privacy (the desire is innate) any more than implosive media create desires for publicity. Instead, they negotiate these desires through amplification and contraction. Understanding how amplifications/contractions may be happening in response to smartphones is among the central purposes of the present investigation. In other words, all people have a desire to control the flow of information they receive and send out into the world. In the following pages, we will investigate what (if any) effect smartphones may be having over these desires, and specifically, if smartphones may be decreasing the level of control individuals have over what information they send out (publicity) and what information they receive (privacy).

Notions of and expectations for privacy have been impacted by the rise of electronic media on a variety of fronts. Tufekci (2008) highlighted how the permanent and searchable nature of online communications make it possible for informal or personal correspondences to be used to publicly shame or embarrass individuals years later and potentially far removed from their original context and meaning. Ibrahim (2008) observed "most social networking sites make it easy for third parties from hackers to government agencies, to access participants' data without the site's direct collaboration, thereby exposing users to risks ranging from identity theft to online and physical stalking and blackmailing" (p.247). Debatin et al. (2009) identify inadvertent disclosure of personal information, damaged reputation due to rumor and gossip, unwanted contact, harassment or stalking, and uses of personal data by third parties as potentially unintended and uncontrollable consequences to individuals participating in online social networks.

Despite the growing concerns related to individuals' privacy, Govani & Pashley (2005) showed Facebook users having low levels of awareness about the company's privacy policies. Their study also failed to find a significant effect from users' awareness of Facebook's privacy options/policy on their utilization of the site's privacy tools. Tufecki's (2008) study showed users' perceived likelihood that their profiles on Facebook and MySpace would be seen by future employers, government agencies, corporations, or romantic partners as not having an impact on their profile's visibility settings. Even as concerns have heightened over the past several years, most users have been shown to be more concerned about which members of their networks, or which "friends/followers" are able to see their information (horizontal audiences) than what information may be gathered by marketers and network operators (Quinn & Epstein, 2018). A recent study by Tsay-Vogel et al. (2018) suggests this may be the result of socialization or individuals mirroring the sharing behaviors of their friends on social networks.

Regardless of users' awareness or concern, the ability of governments or organizations to surveille user behavior is greatly enhanced through mobile technologies (provided they have access to the data mobile devices collect). The discovery of the US Government's collection and storage of most messages communicated though the country's telecommunications networks generated a nationwide debate about the legality and ethics surrounding online data collection (Gellman & Poitras, 2013; Schneier, 2013; Solove, 2011). While this debate continues (Sledge & Watkins, 2015), it is clear that legal and social expectations for privacy are evolving in response to new media technologies.

Privacy Law

Despite its primal/biological basis, privacy is not a right explicitly granted to citizens of the United States. Courts have used the 4th Amendment (prohibition of unreasonable search and seizure) and a number of legislative acts (Privacy Act of 1974, Omnibus Safe Streets & Crime Control Act of 1967 and Electronic Communications Privacy Act of 1986) to create a framework from which a limited legal right to privacy has been interpreted, but the protections are always confined within specific contexts and never absolute. This fragmentation is seen by the number of laws that protect specific forms of personal information (e.g. financial/credit reports, telephone records, health records, motor vehicle registrations, library/video rentals) from being accessed without probable cause by law enforcement and/or other 3rd parties (Banisar, 1999). Generally, four basic privacy protections are recognized today. US citizens are legally protected from (1) intrusions into their private affairs, (2) public disclosures of embarrassing private facts, (3) false media representations (false light), and (4) unauthorized appropriations of one's likeness for commercial purposes (misappropriation) (Sadler, 2005, pp.176-193). For decades, members of the US Congress have been proposing bills aimed at strengthening citizen protections for online privacy, but a strong lobby of business interests in Washington has been successful at maintaining industry self-regulation as the primary means for protecting consumer privacy (Peterson, 2019; Reidenberg, 1999).

While beneficial in the short term for industry, the lack of defined protections is fueling a growing tension between electronic data collection bodies and their users/citizens. At the heart of this tension is the rapid evolution and growing ubiquity of surveillance technologies and legal interpretations of what constitutes a "search" or an unreasonable violation of a citizen's expectation of privacy in the absence of probable cause. Interpretation of 4th Amendment protections has often been done using the "Mosaic Theory" developed by Justice Ginsburg and affirmed by Justice Alito in the US Supreme Court. A central assumption of the theory is that the duration of surveillance/data collection strongly impacts whether a collection constitutes a "search." As the duration of a collection increases, so does the likelihood of it being a search due to the ability to glean greater insights from trends that develop over time. As a simple example, limited inference can be made by knowing an individual visits an office complex once, but if we observe the same individual visiting the same office complex every weekday for a month, it becomes increasingly likely this location is where the individual is employed. In addition to duration considerations, the theory also provides contextual indicators outlining the occurrence of a search happening when information is collected in places where "a person (exhibits) an actual (subjective) expectation of privacy, and (when) the expectation (is) one that society is prepared to accept as 'reasonable'" (Kugler & Strahilevitz, 2015, p. 8).

In their examination of how well this interpretation stands up to modern thinking about privacy in the smartphone era, Kugler & Strahilevitz (2015) show that very few Americans equate privacy expectations with the duration of surveillance. This is not surprising since a singular access to one's smartphone may reveal information that formally would have been stored across a wide range of agencies and media (e.g. travel records, financial data, call logs, Short Message Service (SMS) text messages, personal calendar, and photos). Succinctly, modern smartphones equipped with sophisticated sensors, a variety of application functionalities and high volume storage capabilities remove the need for outside agents to monitor an individual over time. By continually sensing individuals' daily routines, social interactions and content consumption behaviors, smartphones provide unprecedented convenience to users and unprecedented insights about users to anyone able to view and analyze the data contained within them.

Privacy: Contested, Contextual and Commercial

Due to the pervasive surveillance made possible by smartphones and other digital technologies, there was a time when it became a fashionable to say that privacy was a dated and somewhat inconsequential concept (Helen A. S. Popkin, 2010; Sprenger,

1999). However, recent research (the present study included) has pushed back on this view and argues that a better characterization is that privacy has become complex and nuanced in the digital era (P. Kelley, 2021; Kshetri & DeFranco, 2020). In addition to being a biological imperative, privacy is more discussed, debated, and difficult to pin down today than ever before. It is very much alive, evolving and at the center of some of today's most popular and profitable industries.

A starting point for a more thoughtful discussion of privacy is Mulligan et al.'s (2016) argument that privacy should be understood as a "contested concept." Borrowing from the work of Gallie (1955), they define contested concepts as those around which disputes endlessly swirl. Understanding the disputes about such concepts becomes essential to understanding the concepts themselves (e.g., democracy, art, and freedom). Privacy defies a single, unanimous, and universal definition and can likewise be applied in a variety of different contexts: 1) "to determine for (oneself) when, how, and to what extent information about them is communicated to others" (Warren & Brandeis, 1890), 2) "control over where one directs one's attention and how one controls distraction" (Boyle & Greenberg, 2005), 3) "the desire by each of us for physical space where we can be free of interruption, intrusion, embarrassment, or accountability and the attempt to control the time and manner of disclosures of personal information about ourselves (R. E. Smith, 2000), 4) "a threefold guarantee to the subject for awareness of privacy risks imposed by smart things and services surrounding the data subject, individual control over the collection and processing of personal information by the surrounding smart things, (and) awareness and control of subsequent use and dissemination of personal information by those entities to any entity outside the subject's personal control sphere" (Ziegeldorf et al., 2014). These definitions represent a small sample of what can be found with a very general review of the wide range of research on the matter, and while they have aspects of overlap, they also draw attention to different elements of what is encompassed by the term "privacy." Information dissemination is central to Warren and Brandeis, attentional control or focus is highlighted by Boyle and Greenberg, physical and mental solitude is primary for Smith and technological awareness, literacy and control lies at the heart of Ziegeldorf et al. Yet, the mere existence of so many definitions is a strong support for understanding it as a contested concept and one that will mean different things in different contexts to different people. The present study acknowledges the myriad of definitions available, and explores multiple dimensions of the term in the data collection, but keeps at its heart the notion of control over how information is received and sent by individuals (i.e., privacy "as a regulatory process that serves to selectively control access of external stimulation to one's self or the flow of information to others" provided by Klopfer & Rubenstein (1977)) it's guide for measuring evaluating individuals levels of "privacy concern."

Regardless of the definition one chooses to apply, the concept of privacy is contrasted by understandings of publicity. That is, public is the opposite of private. This dichotomy is important when discussing "rights to privacy" in any given context. "Public officials and public figures surrender many of their privacy rights when they enter the public arena" (Sadler, 2005, p. 176). Generally speaking, because public officials (e.g. politicians) and to a lesser extent celebrities are public servants, in the US it is understood that they have fewer privacy protections than private citizens. The effect of which is it is perfectly acceptable for media outlets to report on the criminal records, marital problems, and/or sexual orientations of politicians and celebrities whereas such reporting on private individuals/citizens is illegal.

The internet and smartphones are making this distinction increasingly difficult to identify. To what extent does a private citizen become a public figure when they have a Facebook post that "goes viral?" What are the ramifications of posting the contents of conversation via text message to a website or social network? Should law enforcement agencies be able to access the search engine behaviors of citizens without a warrant? Where this any many other lines get drawn is important, but few public policy decisions have been made to date. Furthermore, it is important to understand if individuals have internalized feelings of public identity as a result of the surveillance and monitoring of smartphone sensors and social media networks may be related to individuals' perceptions of themselves as performing public identities. As such, we will investigate not only how privacy concern levels may impact smartphone use behaviors, but also how smartphone and social media use levels may lead to trading some level of personal privacy in exchange for greater public recognition and/or perceived social importance.

Contextual Privacy

Helen Nissenbaum (2011) has attempted to improve understandings of privacy in digital domains and argues many of the tensions being dealt with online have close counterparts in the physical world. Rather than feeling the need to reinvent the wheel each time a new digital product enters the marketplace, Nissenbaum encourages us to look to social and informational norms that guide and inform similar functions or contexts in non-digital life. For example, protections for consumers and businesses have long existed for advertisements and are regulated by the FTC and FCC. Although online advertising has different capabilities and forms than its predecessors, Nissenbaum's notion of contextual privacy encourages us to look at the norms and practices put in place for print and broadcast advertising as a starting point for the likely expectations and needs for privacy protections online. Or "in our online activity we should look for the contours of familiar social activities and structures… Where correspondences are less obvious, such as consulting a search engine to locate material online, we should consider close analogues based not so much on similarity of action but on similarity of function or purpose" (e.g. library/video rental records) (Nissenbaum, 2011, p. 43).

In addition to examining contextual frameworks and analytical tools to narrow meanings and expectations for privacy, it is also important to consider how mobile platforms/structures impact privacy protections. App marketplaces' (i.e. Apple's App Store and Google's Google Play store) play an important role in protecting and/or limiting users' privacy (Greene & Shilton 2018, Shilton & Greene, 2019). Both Google and Apple serve as the gatekeepers for the apps available to most users of their respective platforms, but the companies take substantially different approaches to decisions about which apps to allow in their stores. Apple maintains a highly restricted app store or marketplace in which app developers need to meet Apple's requirements/definitions for privacy to get their app into mass circulation. Google, on the other hand, takes a laissez-faire app marketplace approach wherein privacy protections are considered features which developers may include or exclude to differentiate themselves from competitors. The authors ultimately found both marketplaces lacking, with Apple's system relying on users' ability to interpret complex and convoluted permission requests (i.e., notice-and-consent) and Google requiring users to have advanced understanding and technical abilities to differentiate between the advantages of different apps and to effectively use the privacy tools different apps may provide as features.

What emerges is an understanding of privacy as a complex, contextual and challenging concept being made even more so by digital technology. As a consequence, it is tempting to become frustrated and potentially apathetic about how one's personal information is being used. Apathy, however, blinds us to very real potentials (both positive and negative) resulting from data collection, analysis, and sharing in the modern media environment.

The Benefits and Costs of Mobile Data Collection

Like the myriad concepts and considerations associated with privacy in general, it is understood that there are a variety of advantages and disadvantages associated with data made possible by smartphones. De Montjoye et al. (2018) present societal improvements that are already resulting from mobile phone data. Among these are improving the management and relocation of populations after disasters, providing real-time traffic alleviation solutions to commuters, and vastly enhancing our understanding and treatment of infectious diseases. Significant economic and financial possibilities are also tied to the personal information collected by smartphones. A new advertising industry that targets audiences based on their behaviors, locations, and demographics has grown up around the data made available by mobile sensors and the monitoring of users online (Acquisti et al., 2016; Danezis et al., 2005). Mental healthcare providers may vastly improve their diagnostic capabilities and patient outcomes while reducing costs by replacing current patient survey data with passive data collections from smartphones (Boonstra et al., 2018). And consumer savings may be reaped from the loyalty programs, targeted ads/discounts and improved healthcare systems that result from these improved understandings and enhancements (Adjerid et al., 2015).

Yet, it is important to realize the same sensors and applications that make these positive uses possible also carry with them the potential to hurt user/societal welfare. Data may be used to identify and target vulnerable communities for scams and thefts, to stalk celebrities or government officials, or to steal users' identities. Algorithms and machine learning systems armed with massive stores of user data may be used for price, job, housing, and education discriminations that disproportionately support the interests of a select few at the expense of most members of society. In other words, data creates a feedback loop which may generate benefits and costs to users. Increases in information disclosure increase the ability of service providers to create more efficient and cost-effective products for consumers. At the same time, and using the same information, service providers can utilize consumers' data to create asymmetric relationships where users may have limited abilities to negotiate (Acquisti et al., 2016), limited understandings of their rights (Ziegeldorf et al., 2014), and limited recourse in the event they feel unjustly manipulated (Banisar, 1999).

Smartphones: The Modern Swiss Army Knife

Two related tools underpin most of the services and controversies related to smartphones. Mobile sensors and internet tracking cookies both provide unprecedented insights into users' behaviors and interests. Understanding how these tools work independently and in conjunction with each other is a first step toward appreciating the potential opportunities and pitfalls surrounding the sharing, sale and analysis of users' personal information.

The modern smartphone contains a wide range of sensor technology that is continually being improved and expanded. Accelerometers, digital compasses, gyroscopes, GPS, microphones, cameras, thermometers, and light detectors represent a sample (but certainly not an exhaustive list) of common sensors found in most smartphones. Used as standalone measurement tools and in combination with one another, these sensors create information utilized by device owners, researchers, businesses and governments (Lane et al., 2010; Sattar et al., 2018). Most of this information is gathered through mobile applications (AKA apps) which are sold and distributed through the previously mentioned app marketplaces. Apps found on most users' smartphones include internet browsers, photography software, navigation tools, social networking portals, news aggregators, calculators, video games, music players and personal calendars (again, not an exhaustive list). The ability to access these functionalities virtually anywhere and at any time provides users with historically unheard-of levels of convenience. Within a single device carried in one's pocket, users can store their book collection, family photos, work presentations/documents, appointments, newsfeeds, credit cards, notebooks, and phone. Additionally, apps can function as personal assistants, health advisors, and educational aids or to facilitate banking transactions, travel recommendations and route navigation. Apps even function to find users romantic partners (Fullwood & Attrill-Smith, 2017). In short, no matter what a user may wish to accomplish, it is very likely "there's an app for that" (Gross, 2010).

In addition to mobile sensors and apps, it is also important to understand the function of internet tracking technology when talking about mobile/digital privacy. Whereas apps may collect user data from a smartphone's sensors and user inputs, internet tracking technology is used to observe user's preferences and behaviors while surfing the web (either through a mobile or desktop web browser). The observation of common internet behaviors (e.g., web searches, purchase histories, web site visits, ad clicks, time spent on site, time spent online, etc.) is most often accomplished by "HTTP cookies."

A deep technical explanation of cookies is beyond the scope of this study (see Kristol, 2001 if interested), but on a general level, cookies are informational storage units that allow web sites to remember past user inputs. This is an important function as cookies improve site functionality, aid user logins, facilitate targeted advertising, and improve webpage load times. As a simple example, cookies allow webpages to remember the items in a user's cart when shopping online. Without cookies, the shopping cart would reset each time a user visited a new product page which would lead to a vastly different (and much less user friendly) shopping experience than what internet shoppers enjoy today. Relatedly, "Flash cookies" serve the same general functions as HTTP cookies but can remember more information and are not saved on a users' web browser (instead stored in the Adobe Flash Player library). Both HTTP and Flash cookies allow service providers and marketing firms to collect data about users, but Flash cookies are usually less understood by users and are much more difficult to remove than HTTP cookies. The increased informational capacity of Flash cookies makes it possible for them to uniquely identify and track users as they "move" around web (often by a Flash cookie field with some variation of userid). The commonly stated rationale for using Flash cookies is their enhanced ability to provide more targeted and theoretically more relevant offers to users, but the enhanced tracking ability also creates potentials for greater invasions of user privacy, especially given users limited understanding of their operation (Soltani et al., 2010).

To reiterate, the point is to give readers a sense of the data and inferential power that results from combining mobile phone sensors, internet tracking technology and user-generated content (e.g., account signups, social/blog posts, product reviews, liking and commenting on content, social networking contacts, and/or virtually any and every internet behavior). The depth of information combined with the ability to identify the users that generate it, allows for the creation of finely detailed portraits of users' personalities, preferences, networks and behaviors. Given this ability and its potential ramifications, an obvious question to ask is why users are willing to provide such unrestricted access to their personal information.

Benefits of Mobile Technology Revisited: Sensors & Tracking

The ability for service providers and data scientists to generate the user portraits described above allows for a variety of benefits. Christin et al. (2011) and Lane et al.

(2010) discuss how mobile devices make it possible to crowd source data collections that are currently labor and resource intensive to perform. For example, mobile devices make it possible to task individuals with pollution monitoring which can lead to targeted efforts at reducing carbon footprints, tracking the spread of disease to lessen the effects of outbreaks, and monitoring animal migration patterns to improve wildlife preservation strategies. Similarly, Guo et al. (2018) envision future interactions between mobile technology and physical environments that could revolutionize daily life. Examples they forecast in the relatively near future include smart roads interacting with mobile sensors to form vehicular networks that will reduce or eliminate traffic congestion, education software that uses mobile devices to interact with objects in students' physical surroundings to turn everyday environments into immersive interactive classrooms, and health monitoring sensors which identify symptoms of disease before users perceive them and notify the appropriate professionals within their health networks.

Smartphones may also revolutionize mental health treatments (Beierle et al., 2018; Matteo et al., 2018). Observations from mobile sensors combined with monitoring of internet browsing behaviors make it possible to infer users' emotional states, sleep patterns, and levels of social interaction. All of which are useful for identifying symptoms of depression, anxiety, and aggression in addition to providing insight about patients' coping strategies. Boonstra et al. (2018) highlight the potential for passive data collection through mobile devices to replace current methodologies that rely on patient provided survey responses. This passive collection from mobile devices has the potential to provide more detailed (e.g., daily patterns), efficient (cost effective) and less biased data than self-reported patient responses which rely on recall and diary entries.

Ippoliti & L'Engle (2017) showed mobile phones could be ideal mediums for communicating healthy living practices to individuals with low-incomes and those living in areas which are difficult for aid workers to access. Text messages were demonstrated to be efficient and effective at communicating information related to sexual health to youth in areas being overridden by sexually transmitted disease. The messages were effective at improving health knowledge, reducing sexual risk behavior and increasing utilization of local health services. Indeed, the ability to better collect, store and share health information with patients is often cited as one of the key benefits being made possible by mobile and digital technologies (Acquisti et al., 2016; Adjerid et al., 2015; Papageorgiou et al., 2018). In short, smartphone data has the potential to help doctors make more informed diagnoses and treatment plans that improve patients' standard of living and save lives.

Research has suggested that poverty reduction efforts may be improved by adopting mobile technology. Steele et al. (2017) analyzed mobile phone data, movement patterns and social networks inferred by mobile communications to identify geographic pockets of poverty in areas of the world where such information is generally unavailable due to insufficient funding. In addition to mapping poverty in areas typically unmeasured, the authors point out that as "sensing data become more widely available, analyzing (poverty) data at regular intervals could allow for dynamic poverty mapping" (p.8). Such mapping may save time and resources which could be repurposed for relief and reinvestment initiatives. Relatedly, Björkegren & Grissen (2018) devised a method for predicting default rates based on mobile phone usage data. In many parts of the world, formal credit reporting is not possible due to there not being credit bureaus that monitor credit for individuals living in low-income areas. With a relatively small amount of mobile phone data, the authors created "a method to predict default among borrowers without formal financial histories… (which) could conceivably be used to predict other outcomes of interest – such as lifetime customer value, or the social impact of a loan" (p. 22).

Thus, conversations that center on user benefits related to targeted/relevant advertising, convenience of digital libraries, and accessibility to others are accurate, but incomplete. There are massive potentials for smart phones and other mobile communication devices to improve health, transportation, education, environmental and financial outcomes for individuals throughout the globe. Yet these benefits come with a cost that could reduce privacy and ultimately autonomy for users and societies at large.

Potential Negative Costs of Mobile Data

With enough information, it becomes a very real possibility for an agent to control the thoughts and behaviors of another (André et al., 2018; Yeung, 2017). As will be discussed in detail in a subsequent section, armed with a detailed purchase history, it is possible to infer a potential customers' reservation price (or the maximum price a customer is willing to pay for a specific product or service) and consequently render an economic negotiation meaningless. Extrapolating this concept, with enough information about one's political beliefs, religious preferences, personal/professional networks, media consumption behaviors, level of education and mental/physical health, the ability to manipulate actions, feelings and thoughts increases significantly. Indeed, this potential is at the heart of concerns related to Russia interfering in US elections (Rathi, 2019; Ribeiro et al., 2019; Valle et al., 2018) and will only become more pronounced as sensor and internet tracking technology expands in use and increases in sophistication.

In the US, users are tasked with protecting themselves from these abuses (President's Council of Advisors on Science and Technology, 2014). Providing users with privacy policies and app permission requests that describe how information is collected, used, and shared has been deemed an adequate means for giving users the ability to make informed decisions about protecting themselves from unwanted use. This process is commonly referred to as, "notice-and-consent" and is used by virtually all websites, mobile application, and smartphone providers. There is ample evidence to suggest that users do not possess the ability to make these decisions (discussed shortly). Nonetheless, several efforts have been made to improve users' abilities to make these decisions or to better align the use of personal data with users' stated desires for privacy. These efforts take a variety of forms and include highlighting specific elements of privacy policies or providing policies at opportune moments, translating permissions into more user-friendly terminology and creating technological protections that obscure or encrypt user information.

Aiding User Choice & Protecting Personal Information

The collision of complex, integrated and highly technical data collection devices (i.e., smartphones) with an expectation for users to manage how these devices operate generates the need to understand how users make decisions about sharing their information. In an effort to aid users' understandings of privacy policies and app permissions, a common approach is to improve the layout and word choices of policies to facilitate greater engagement and comprehension. Schaub et al. (2015) developed app permission requests that tailored the timing, channel, and modality of the permissions to maximize user relevance, comprehension, and recall. By improving the information contained in Android permission screens, Lin et al. (2012) increased participants' understanding of the requests and often lowered concerns about using apps. Policies that were streamlined to make information sharing policies more salient and comprehendible led participants in Liccardi et al.'s (2014) study to select more protective and secure apps. Yet a theme throughout all these studies was finding that absent the improvements/modifications provided by the researcher's, users were generally unaware of how apps may use their information.

Independent of users' ability to comprehend policies, Acquisti et al. (2017) cite a variety of biases recognized in the economic literature (e.g. hyperbolic time discounting, optimism bias, availability heuristic/bounded rationality) that may undercut rational app selections/decisions. In response, they propose design elements that may directly reduce user bias. For example, "risky" settings could be made difficult to configure (in their words, cognitively costly), feedback about app information sharing could be provided before *and during* system use (to alert users about sharing they have authorized when apps are in use), and users could be required to perform a cognitive test or calming exercise prior to agreeing to risky policies in order to avoid "hot state" (i.e., emotionally charged) choices.

In addition to improving understandings of permission policies, researchers have developed tools that serve as aids and provide recommendations to users based on their desired levels of privacy. Tsai et al. (2010) conducted an experiment to see if users were willing to pay a premium for privacy by giving them a shopping task aided by a privacy tool that displayed graphics to differentiate information about retailers' privacy policies. Their results showed that participants were willing to pay a small premium for privacy but were not willing to pay higher premiums for more sensitive purchases. Nonetheless, given access to a tool that made privacy policies accessible, easily interpreted and differentiated, users did show a willingness to pay a premium to purchase from retailers that provided some protections for their personal information collected during the purchasing process.

Perhaps the most ambitious privacy tool seen in the literature is Liu et al.'s (2016) development of a "personal privacy assistant." Instead of aiding users' management of permissions on an app-by-app basis, their app functioned as a master control center that could recommend and implement privacy settings within every app a user had installed on their device. The personal privacy assistant collected responses from users about their desired levels of privacy protection in different contexts to create a privacy profile. The profile was then used to generate recommended settings for each app a user had downloaded to their smartphone. Most users were placed in the least restrictive profile (i.e., the profile that allowed the highest level of sharing) and those that were not initially assigned to this group often chose to be moved into this group due to apps not functioning correctly under more restrictive settings. Their discussion highlighted two findings. Primarily, their tool showed the potential for significant gains in efficiency when a single app is given charge over managing permissions across a device (rather than users being tasked with managing apps individually). Secondly, their findings illustrated how most smartphone applications utilize user information for multiple purposes and do not facilitate the ability for users to allow sharing for one function and not another (e.g., to allow for an app to use one's location for navigation purposes but not for advertising).

Safeguarding Privacy with Technology

The complexity of permissions and level of sophistication users need to effectively manage their requests has led to the development of technological protections designed to protect users' information regardless of their or a company's abilities to protect it. In order to understand how information can be maliciously obtained, Toch et al. (2012) identified three risk points: data collection, model creation and the adaption of a model to uses (e.g. sharing user insights/profiles with users themselves or with other interested parties). In reviewing the literature, they identify a variety of methods which can be used to alleviate the potential for abuse at all points. Distributing data across multiple storage locations (preferably all owned by users) prevents an individual site that is compromised from sharing the totality of a users' information. Obscuring data after it has been collected is another potential remedy. Perturbation is a process of adding "noise" to data by adding random values to a set before it is sent to outside parties (p. 211) and similarly, obfuscation replaces a specified percentage of collected data with random values. In both cases, users can "hide" behind the potential for any element of the models coming from random, rather than individual/unique values.

These techniques in combination with user controls/tools may help prevent unwanted uses of user information that could limit autonomy and/or violate users' desired levels of privacy protections while simultaneously improving profitability for companies. Apple recently began using a form of perturbation/obfuscation in its products known as "differential privacy" (Simonite, 2016). In doing so, the company believes they have prevented the possibility of users being uniquely associated/identified with data collected about them (i.e., ensuring user anonymity), and users are still free to tailor their app privacy settings to reflect what information gets collected. In other words, technological data protections may enhance user privacy and provide companies with competitive points of differentiation in the marketplace.

While these developments are encouraging, every generation of smartphone presents new challenges for protecting users' information. In response to the growing sophistication of sensor technologies (specifically, improved mobile analytics and facial recognition technology), Osia et al. (2020) modeled a "deep neural network" that extracts relevant information for a specific function or app feature on a user's device before sending it to cloud servers for processing and service delivery. These kinds of extractions may become increasingly necessary as sensors capture more and more superfluous data (e.g., Siri/OK Google on smartphones and smart home devices like Amazon Echo and Google Home that "always pick up" audio and video data to respond to user requests). In addition to lowering the computational load on a device which would be needed for encrypting this volume of information (regardless of the various possible methods: k-anonymity, differential privacy, noise addition, etc.), feature (i.e., relevant data) extraction makes secondary/non-core use of data much more difficult to collect and theoretically prevents uses of data outside what users would reasonably expect to receive a desired service.

In total, there are many tools and technologies that have been designed to aid users in managing how their data is collected and used by third parties. This study and all of those above reflect a considerable effort on the part of many researchers and some companies to protect users' information and address users' desires for privacy. Serious attempts to improve how privacy policies are communicated, how users make app selections and how applications collect and transmit data have been and continue to be prevalent topics of concern. However, very few of these tools and technologies are currently available to the public and many social media platforms' business models are built upon harvesting and selling user data. The tension between these efforts will increase as threats to users' information are continually evolving with each advance in mobile/sensor technology.

Privacy Threats: Malware, Unexpected Use & Big Data

Despite the efforts of researchers and many companies, smartphones will always have the potential to compromise their owner's privacy. There is a growing consensus that full anonymity or complete privacy may never be possible for smartphone users (Narayanan & Shmatikov, 2009; Polakis et al., 2015; President's Council of Advisors on Science and Technology, 2014; Ziegeldorf et al., 2014). Although specific threats vary in functionality and purpose, privacy threats usually come about through one of three modalities. Applications may be specifically designed to attack users and collect their information without their knowledge. Users may permit the collection of information by an app, but have it used in ways the user does not anticipate or that was not possible at the time it was collected. Finally, data collected from many different apps may be aggregated into a single unifying piece of information or insight that would not be possible by having the individual data points in isolation from one another.

The most straightforward of these is mobile malware or malicious apps that are designed to confuse users about their true purpose. Felt et al. (2011) provide a good overview of how mobile malware is being utilized and distinguishes three categories of threats: malware, grayware, and personal spyware. Malware steals user information or incapacitates the device in some fashion (e.g., trojans, worms, botnets, and viruses). Personal spyware takes the control of a devise away from the user to download illicit software without the user's knowledge or consent. Grayware are legitimate apps that provide users with a promised service, but also collect user data for purposes other than the app's primary function (usually for profiling and advertising sales).

Investigating how much malware was available in the popular app marketplaces revealed Apple did a better job than Google at preventing malware and personal spyware. However, grayware was found in all the major app marketplaces and users who "jailbroke" (Apple) or "rooted" (Google) their devices faced far greater risks of downloading malware and personal spyware. That is, applications only found outside of the sanctioned marketplaces were much more likely to be or to contain malware for both Google and Apple devices.

To some extent, users may protect themselves from malware and even grayware by thoroughly reading and understanding app permission requests prior to downloading new apps (easier said than done). However, several researchers have demonstrated how seemingly mundane data can be used for surprising and unexpected purposes. Legitimate applications may be attacked/hacked as well, at which point their permissions or protections for users become obsolete.

Michalevsky et al. (2015) were able to use a phone's power consumption rate to infer users' physical locations. Although only possible when users were traveling quickly (by car or bus), their findings nonetheless demonstrate how permissions users may interpret as benign (in this case, access to power/battery consumption which would not have required a users' consent at the time of the study), can be used to collect data for purposes beyond what a careful and concerned user would expect after reading an app's permission requests. Similarly, Alepis & Patsakis (2017) used WiFi-P2P data on Android devices to observe users' locations. WiFi-P2P is an Android feature that allows devices to link together without the use of Bluetooth. App access to this functionality does not require user permission, nor do users get notified when WiFi-P2P is in use. In other words, unintended consequences to user privacy may result from software/hardware design decisions that are only uncovered long after devices are well integrated into the marketplace.

Moving beyond permissions (or the lack thereof), Polakis et al. (2015) exposed vulnerabilities in the most popular social networking apps which allowed them to reveal users' locations with a high degree of precision. They were able to pinpoint stationary Facebook users within five meters of their exact location and had similar results for every major social networking application. Thus, even when application designers can be trusted with sensitive information, there is always the potential for adversaries to gain access to it, and once had, permissions and often even legal protections become moot.

There is also evidence that users tend to misunderstand or simply do not notice alerts designed to inform them of surveillance taking place. Participants in Portnoff et al.'s (2015) study were mostly unaware of being monitored by a computer's camera despite the illumination of a LED next to the camera while it was on. Of those who did notice the LED, almost half may not have understood that its illumination meant the camera was recording. On mobile devices, indicators of sensor activity are even smaller (or non-existent) and are likely to have similar levels of understanding. Going a step farther, if a smartphone were compromised, these findings indicate many users may be oblivious to their phones acting in ways they likely would not want and to which they had never consented.

Lack of awareness and understanding carries with it the potential for serious consequences. Spensky et al. (2016) argue the complexity and interconnectedness of mobile devices is what makes them simultaneously highly functional and extremely vulnerable. For users, it is convenient for a picture library to able to "communicate" with text messaging, email, and social networking apps to make photo sharing a user-friendly experience and/or for location sensors to link with weather applications for more accurate/relevant forecasting. Yet, this same interconnectivity of sensors and applications means when a single sensor or app is compromised, it can lead to the compromising of security for the entire device.

In summary, the connectivity of apps which facilitate internet browsing, phone calls, text messages, banking, navigation, social networking, and content consumption with sensors that identify images, faces, objects and can measure mental health indicators, physical activity levels, travel behaviors and environmental factors generates unprecedented potentials for consumer services and surveillance. Users stand to benefit greatly from enhanced knowledge of themselves, their communities, and their environment with attendant risks to the most sensitive details of their private lives. Collectively, the collection, analysis and sharing of these details through cookies, flash cookies and mobile apps has given rise to an industry known as big data.

There is little debate that the greatest threat to privacy comes from the power of big data companies and organizations. Much like privacy itself, definitions of big data are numerous, but a concise view was provided by the President's Council of Advisors on Science and Technology (2014) who state, "big data typically means data about one or a group of individuals that might be analyzed to make inferences about individuals" (p. 2). As has been discussed, the ability for mobile device sensors and internet tracking cookies to collect information is robust and continually being improved and expanded. Likewise, methods for analyzing and transforming this data into user profiles/inferences are also rapidly evolving. Recognition, sorting and pattern identification algorithms are at the heart of new sciences for machine learning and artificial intelligence which are fundamentally changing approaches to manufacturing, advertising, education, agriculture, healthcare and entertainment (Kietzmann et al., 2018; Makridakis, 2017). The distinct firms that collect and transform big data into a product for consumption are collectively referred to as the Big Data Industry (K. E. Martin, 2015). The role of these firms is essential to distilling the huge volume of information being collected by computing technologies into "actionable information". By aggregating and filtering big data, these firms create new data sets which are generally tailored for marketing professionals in the wide range of industries listed above.

While the potential for data and advanced algorithms to benefit society on virtually every level are exciting, this power also creates unique challenges. Central to the discussion is the ability for algorithms to "crack" encryption and other anonymity preserving methods discussed above (Chester, 2012; Ghahramani, 2015; Narayanan & Shmatikov, 2010; Wind et al., 2016). Big data results in an environment where any single data point known about a user (no matter how benign it may seem on the surface) can become a means to uniquely identify the user. The vast stores of user data collected by ad networks, social networking sites, mobile apps and internet tracking cookies make it virtually impossible for individuals to avoid having their data collected in large volumes. With the support of vast libraries of user information, any single data point can become a meaningful identifier.

Illustrations of individual data points leading to larger inferences are now common. Narayanan & Shmatikov (2009) were able to accurately de-anonymize networks containing thousands of individuals with just one user's username and a couple of common characteristics publicly displayed on social networking profile pages (i.e. one weak link in an entire social network compromised everyone within it). Botta & Genio (2017) used mobile phone call records to infer social networks and community behaviors with a high degree of accuracy (i.e., strength of relationships, community structures, and patterns of daily life). And, as discussed, locational and travel behaviors have been observed using data from seemingly unrelated mobile sensors (e.g., battery consumption).

The potential and power of user inferences from big data will continue to grow as mobile technology and internet tracking technologies become increasingly sophisticated and as new data becomes available from the adoption of new devices (e.g. Internet-of-Things, see Wortmann & Flüchter, 2015). Data scientists will continue to hone and develop algorithms to process big data for the purpose of improving services related to users' health, entertainment, work, education, and travel (among many others). At the same time, the ability for layman users of these technologies to understand the complexity and connectivity of back-end processes related to data collection, analysis, profiling, and service delivery will continue to decrease. In many ways, the smartphone and digital communication landscape is increasingly coming to mirror the relationship of many car owners with their vehicles. Most lack the ability to understand the mechanics of their vehicles, and so too is the complexity of digital technologies making it impossible for most individuals to comprehend the full scale of how their personal information is being collected and used by their smartphones.

Failure of Notice and Consent: Information A-Symmetry, Design, Cognitive Bias & Complexity

Despite the growing complexity of mobile and internet technologies, the onus of responsibility for protecting one's information in the US currently resides with users. Transparency or providing users with privacy policies on websites and permission requests on mobile apps has to this point been deemed an adequate empowering of users to make informed decisions about information sharing. That is, a "notice-and-consent" framework which asks users to agree to various data practices prior to using different applications is how users' data is protected in the US.

Giving users the ability to decide for themselves how their data should be used is an appealing concept but creates a number of difficulties in practice. There is a growing consensus that asking users to make informed decisions in the era of big data and pervasive sensor technologies is unrealistic (Adjerid, Samat, et al., 2016; Athey et al., 2017; Landau, 2015; Liccardi et al., 2014; A. M. McDonald et al., 2009; President's Council of Advisors on Science and Technology, 2014). Even if users possessed firm understandings of their smartphone's ability to collect information about them (a high and unlikely bar to clear), the complexity of data sharing relationships between ad networks, social networking sites, search engine operators (e.g., Google) and governments is often beyond the ability of most individuals working in these industries to comprehend, let alone non-professional users. Furthermore, these relationships are almost always obscured (if not completely hidden) from users. Users often agree to a service's concisely stated policy of sharing information with third parties, but it is very unlikely for them to realize how this agreement may lead to using their information in ways that would greatly violate their desires for privacy.

One of the biggest hurdles to users being able to make informed decisions comes from the language used in most privacy policies. Even expert readers (i.e. legal scholars/academics) in Reidenberg et al.'s (2015) study were unable to agree upon interpretations of privacy policies for some of the most popular websites. Not surprisingly, knowledgeable (grad students in relevant fields) and layman users were similarly confused. The overwhelming conclusion was the vast majority of policies are poorly communicated, "experts could not reach consensus on interpretation of data sharing practices generally and agreed even less as to the nuances of data sharing" (p. 46). McDonald et al. (2009) had similar findings when investigating the reading levels required for comprehending common policies. In summarizing their results, they found "even the most readable policies… too difficult for most people to understand them and even the best policies… confusing" (p. 51). These results make it easier to understand why many users do not bother reading the policies at all (Jensen et al., 2005).

Design, cognitive bias and time constraints have also been shown to impact sharing decisions and app choices. Acquisti et al. (2015) summarized a list of contextual factors that may alter an individual's sharing behaviors including convenience/default settings, perceived control over audience, behaviors of friends/others, and the potential for gaining social capital from publicity. Decisions made in hypothetical versus real world settings have also led to different choices. Adjerid, Peer, et al. (2016) found that users gave more credence to objective changes in privacy policies/protections in hypothetical situations than they did in actual settings. That is, actual/objective risk increases were a greater influence on hypothetical behaviors than they were on actual behaviors which were governed more by relative improvements (i.e. if a policy/behavior became relatively more protective, even if still objectively very risky). Furthering these findings, Adjerid, Samat, et al. (2016) showed differences in how sharing decisions were framed impacted user choices. When given an acceptance frame (i.e. will you allow sharing), users were significantly more likely to provide a permission than when given a reject frame (i.e. do you prohibit sharing).

The impact of notification design was shown by Egelman et al.'s (2013) finding that presenting users with side-by-side comparisons of app permissions (not possible in the current app marketplaces) led to many users making app selections more aligned with their stated desires for privacy. In contrast, users making decisions in a simulation mirroring current marketplaces (with a single app requesting a list of different permissions), were less likely to stick to their stated desires for privacy. In total, research has shown privacy policies to be incomprehensible and users being easily influenced by the framing and design choices designers make about communicating their policies and practices.

A final note concerning the inadequacy of the notice-and-choice framework to institute user preferences is found in users' inability to use most tools provided for protecting their privacy. Leon et al. (2012) examined several of the most popular tools designed for protecting users online (Ghostery, TACO/Targeted Advertising Cookie Opt-Out, AdBlock Plus and Internet Explorer Tracking Protection) and found them to be universally unusable by users. The inability for users to distinguish between trackers, inappropriateness of default settings, communication issues, need for feedback, use of confusing interfaces and need to not break websites were major issues preventing effective implementation of all the tools evaluated. While impossible to prove, the authors speculated the ineffectiveness of these tools may be intentional since most free services and apps are supported by the selling of user data. Generally, all of these findings illustrate a mobile/digital ecosystem that is highly complex and interconnected which only the most expert users can begin to understand and attempt to manage.

The unrealistic expectation of users to successfully manage information sharing in the big data era has led to a growing outcry for notice-and-consent policies to be replaced by a governing structure that handicaps users' ability to make informed decisions about their how their personal information is used. Rather than giving users the responsibility of managing their data, a number of experts have advised making data users (i.e. marketers, social networking site operators, searcher engine providers and governments) responsible for protecting data and respecting users desires for privacy (Acquisti et al., 2015; Landau, 2015; President's Council of Advisors on Science and Technology, 2014). In addition to these being the only entities with the time and resources to address the complexity of big data, giving these entities explicit responsibility may also help to remedy data breaches and vulnerabilities that have been increasingly common over the past several years (Berghel, 2017; Manworren et al., 2016; Papageorgiou et al., 2018; Ristenpart et al., 2009; Trautman & Ormerod, 2016).

What Users Want: Understanding Users Conflicting Desires

To create frameworks by which data firms or any other body will protect and honor users' privacy expectations, it is necessary to understand the privacy preferences and desires of users. Given the contested nature of privacy as a term and the complexity surrounding data collection in the pervasive computing era, it should come as no surprise that users' preferences are often as nuanced and complex as the big data industry itself. On a macro level, most Americans express a high level of concern about how their personal information is being collected and used. "Some 81% of the public say that the potential risks they face because of data collection by companies outweigh the benefits, and 66% say the same about government data collection" (Auxier et al., 2019). However, there is also ample evidence to suggest that privacy concerns are mitigated, confused, or ignored within varying contextual frames.

One of the most interesting examples of users sacrificing privacy is observed when they are given the ability to control which members of their social circles can see the information they share. Brandimarte et al. (2013) demonstrated the counterintuitive finding that providing individuals with greater control over which members of their intended audiences could see their content, led to riskier sharing decisions. That is, by giving individuals more options about which known others in their networks were able to see their postings, authors were able to induce participants to share more sensitive personal information and to share this information more frequently. This tendency supports conclusions that individuals simply do not understand or consider the big data industry when choosing to participate in many online behaviors. "People fail to distinguish between two importantly different forms of control… control over *release* of personal information (the action of willingly sharing some private information with a set of recipients) and control over *access* and *usage* of the information by others (because the final audience who has access to released information may include third parties who gained indirect access to it, and even intended recipients may use the information in unpredictable ways" (p. 343).

Relatedly, Benisch et al. (2011) showed location sharing increased when users were given more fine-tuned controls over when and where their locational information was shared with others. The authors made the logical argument that users erred on the safe side (i.e., did not permit sharing) in the absence of control, but gradually choose to share more as their level of control increases (i.e., as they gained control over which members of their audiences would see what they shared). In both studies, users seem to exclusively focus on members of their intended audience (family, friends, coworkers, etc.) while completely discounting or ignoring how their sharing may be monitored by cookies, ad networks, device manufactures, and the operators of social networks (i.e., big data entities). Furthermore, these examples illustrate the "control paradox" wherein users make risker decisions about sharing personal information as their perceived level of control over which members of their intended audiences will be able to see it.

If users were more cognizant of how third parties access their information, there is evidence to suggest they would be more wary of sharing. Xu et al. (2009) showed that users were generally more comfortable with their locational data being used by companies they had opted-into receiving marketing communications from (i.e., using pull tactics), but were much less comfortable when their location was shared for marketing purposes with unknown firms (i.e., push tactics). This finding supports the idea that users may engage in a "privacy calculus" where a certain level of privacy loss is acceptable in exchange for desired services/offers, but the equation breaks down when users are unfamiliar with a firm with whom they are sharing since they are incapable of evaluating the value of the services/offers it may be able to provide in exchange.

Contextual elements of decision making are also seen by Martin & Shilton's (2016) examination of different application functions. Responding to "factorial vignettes," participants indicated far more acceptance of sharing their location information for weather and navigation than for the purposes of gaming and social media (p. 210). Bilogrevic & Ortlieb (2016) had similar results showing that the type of service (banking, search engine, social network), type of data (location, financial info, contracts), and relationships with third parties (e.g., ad networks and partner companies) were all impactful on participants comfort levels with their information being shared.) Thus, general surveys measuring privacy concern may overlook the nuances of different app functionalities and user tolerances for sharing in return for different services, but in all cases where information is shared with unknown third parties it is likely to violate user preferences.

Users have also shown far less logical responses to sharing decisions. Patil et al. (2014 and 2015) showed mismatches between users' acceptance of information sharing depending on if they were notified in real-time or after the fact. Informing participants their location had been shared in real time generated greater negative reactions than waiting for a designated disclosure time each day. Additionally, predefined preferences were found to be overly restrictive (i.e., predefined rules prohibited sharing in instances where users later identified it being acceptable to share) when compared to real time requests to share. Both studies showed that timing and users emotional states have meaningful impacts on expectations and desires for privacy which ultimately impact decision making.

It is also important to note participants generally perceive greater benefits going to companies as a result of their information being shared than they receive in exchange (Bilogrevic & Ortlieb, 2016, p. 5224). This imbalance is likely to become increasingly relevant as more and more sensor technologies enter individuals' homes and workplaces. Naeini et al. (2017) support the view that perceived benefits will be a key determinant in users' acceptance of Internet-of-Things (IoT) technologies. They identify duration of information storage, where information is collected (i.e., public vs. private spaces), and data type (e.g., biometric vs. social) as influential considerations in IoT contexts. They also circle back to a theme already seen regarding users general misunderstanding of the power of big data to make inferences. They point out participants indicated greater comfort with "anonymous" collection but were seemingly unaware of how IoT will make re-identifying "anonymous" data relatively easy (p. 409).

While contextual in nature, the discussion of users' desire for privacy underscores the fact that most feel they are getting the short end of the stick when it comes to information sharing online and through mobile applications. Despite industry arguments that users benefit from convenience and more useful marketing communications, Turow et al. (2009) and A. McDonald & Cranor (2010) showed majorities of Americans across all age groups do not desire targeted advertising. Facebook was forced to scrape their "Beacon" initiative that shared users purchasing decisions with others in their networks without prior consent (Story & Stone, 2007). Designed as a tool that would provide "relevant" information to similar consumers, user backlash to the program quickly made defense of the tool untenable. These examples illustrate that many of the theoretical benefits that result from the collection of consumer information do not deliver much value when evaluated by consumers themselves.

These studies also showed most participants having un or misinformed beliefs about how behavioral advertising audiences were curated and what legal protections are in place for their personal information. Participants often believed in legal protections that do not exist and underestimated how much information about them was being collected and used for advertising purposes). Consequently, there could be even greater dissatisfaction if users were more informed. This is not surprising since the data collected by app and cookie tracking happens in background processes to avoid interrupting users' activities, but as a side effect, users are typically unaware of when their information is being collected and even less aware of when their information is shared or sold to third parties. When a user sees a product of these practices (e.g., when an advertisement for the car they just searched for appears in their newsfeed or when a promotional offer for a nutrition program appears after signing up for a gym membership) it is very likely they can infer what has happened at a general level (their behaviors are being communicated to advertisers), but how to take control of their personal information is often beyond their understanding and agency.

The desire for privacy, but inability to manage one's information is a salient issue in the literature and has been growing as sensor/mobile technology has increased in sophistication. Boyles et al. (2012) found 54% of app users had decided not to install an app after realizing how much information it was collecting. They also found "50% of smartphone owners (having) cleared their phones browsing or search history, while 30% (had) turned off the location tracking feature on their phone due to concerns over who might access that information" (p. 8). That is, many American's were trying to address their privacy concerns with proactive actions they believed were effective. By 2019 however, Auxier et al. (2019) found over "80% of Americans indicating they had little or no control over the data collected by the government or companies (p.6). 70% reported their information being less secure than it was in 2014 (p. 15) and 72% believing all or most of what they do online or on their cellphone was being tracked by companies" (p. 25).

The perceived lack of control illustrated by these statistics may also help to explain some previous researchers' findings that users discount privacy permissions when compared to other more easily understood and actionable app characteristics. Kelley et al. (2013) found users rated price, recommendations, design/aesthetics, and convenience above privacy protections when making downloading decisions. Yet, users also indicated they felt their information was readily available already and that they had little control over how their personal information was collected and used. When given tools that made comparison of privacy permissions easy to make, many users did consider this information and would often choose applications with fewer permissions provided the apps provided the same level of service. In other words, these findings may be less about users not caring about privacy and more about them feeling like it has already been compromised. With this mindset, they feel helpless and turn to considerations over which they perceive more control (e.g., how much they spend, which interface they like best, storage requirements, etc.).

Of course, individuals making decisions that run counter to their interests is not a new phenomenon. Originally designed to test economic theories of rational choice, a large body of research has observed behavioral/emotional factors that lead to irrational decisions. Loewenstein's (2000) work argued that despite being necessary for survival and quality of life, passions/emotions "such as anger, fear and embarrassment often cause (people) to act contrary to their own economic interests" (p. 430). Ajzen et al. (2004) observed behavior often deviating strongly between what individuals *say* they will do in each situation and how they *act* once placed in that situation. Central to their findings were societal norms influencing what respondents think is the "right" behavior in hypothetical tasks, versus more selfish considerations when making the choice in real world contexts (they also show this bias can be corrected for by effective priming in hypothetical situations that put respondents in a more "real-world" mindset). Providing an illustrative example, every participant in FeldmanHall et al.'s (2012) experiment was willing to subject "fellow participants" to painful shocks in return for financial compensation despite only seven percent saying they would keep compensation at the expense of other participants prior to the start of the experiment.

These economic findings have been tested as explanations of the gap between *stated* privacy concerns and actual information sharing *behaviors* in online settings. Keith, Babb, et al. (2014) designed a "geo-caching" game to compare location sharing decisions while playing the game to survey responses measuring privacy concerns prior to playing the game. In discussing their findings, they found participants exhibiting "bounded rationality" or being more strongly impacted by relative changes in risk positioning than objective threats to their information. Similar findings by Peer & Acquisti (2016) attributed bounded rationality as the explanation for "reversibility" impacting participant decisions. Counter-intuitively, when users were told they would have the ability to revise or delete information they posted, they provided less information than participants in a condition that did not allow them to revise/delete their postings. They believed this was due to information sensitivity becoming more salient when users were primed to think about revising/deleting their responses as opposed to simply being asked to respond to questions or supply information. In both cases, rather than using their stated personal preferences for privacy as a benchmark, users seem to readily discount their initially stated desires and instead focus on factors salient within their immediate contexts.

These findings illustrate a clear disconnect between behavior and stated desires for privacy protections. Barnes (2006) was the first to coin the term, "privacy paradox" to describe this gap when she observed how youth readily engage in posting information to their social media accounts that can be used to develop profiles for targeted ad campaigns, identify their place of residence, and more generally monitor their activities for beneficial and nefarious purposes. Adults, by contrast, expressed strong opposition to third parties having access to these kinds of information (and about their children), but did little to curtail their social media behaviors (likely due to not understanding the capabilities of these networks at the time of the study). Norberg et al. (2007) subsequently formalized thinking of the "privacy paradox" and drew focus to the complexity and enormity of big data collections, analysis, and use. They were among the first to identify the information asymmetry users faced regarding decision making about privacy policies and permissions. The sheer volume of information and myriad analysis tools available to big data entities makes it extremely difficult for users to make informed decisions about which apps to download and what information to share. The paradox continues to be readily observed in the literature, with a recent study showing users choosing to surrender privacy in exchange for seemingly small rewards, convenience and/or entertainment despite their concerns about how their information is used (Athey et al., 2017).

Presence & Place | Attention: Cognitive Resources & Interruptions

The discussion of privacy has illustrated the power of smartphone data to generate insights about users' cognitive and behavioral characteristics. A central point being made by many of these studies is that user analysis/profiling will happen regardless of a user's awareness of or engagement with these practices. Whereas data collection generally happens outside of an application's core functionality and users' notice, an application's notifications are designed to solicit users' attention and prompt them to engage with their smartphones. Importantly, this means smartphones both stand at the ready for users to make use of them and ask the users to make use of them through notification alerts. Alternatively, the ability for users to always *access* information and contacts in all places also always makes them *accessible* to their apps and contacts. This increased access and accessibility has been shown to impact users' cognitive functioning, behaviors, and ability to perform tasks in settings that increasingly blur the boundaries between physical and digital space.

Environments in which individuals are present both physically and digitally (e.g., when one responds to a text message while attending a performance or sporting event) generate interesting questions about how smartphones affect attention and focus. It is useful to think about attention as "the allocation of mental resources, (which is) hinged to different levels of (mental) workload" (Patten et al., 2004, p. 342). Central to the definition is understanding attention as a finite resource. Different individuals may have varying capacities and management capabilities for it, but everyone has a limit to how many tasks or mental processes they are able to engage with at any given moment.

As a limited resource, the ability for users to allocate attention between smartphones and their physical environment has been a popular topic of study. As it has serious safety ramifications, numerous researchers have investigated the impacts of mobile phone use while driving which continues to the present day (Alm & Nilsson, 1995; C. Liu et al., 2019; Reimer, 2009; Trbovich & Harbluk, 2003). As most readers will know, the results show smartphone use while driving drastically increases the probability for severe collisions or that smartphones seriously hinder individuals' ability to perform well on driving tasks. Yet, driving is far from the only activity impacted by smartphones. How an individual allocates their attention is a process that plays out at every moment of life and as smartphones increasingly permeate users' daily lives, so too does it impact the way they allocate their attention to others, tasks and objects in their environment.

To begin, it is important to acknowledge the nearly infinite variety of content and communication opportunities smartphone users have at their disposal. The ability for users to carry entire catalogs of motion pictures, YouTube videos, music, video games, news sources and literature (to name a few) in their pocket gives them on-demand access to more information than an individual can consume in a lifetime (or several lifetimes). The consequence of having what is in effect infinite content choice is a battle for consumers eyes and ears known as the "attention economy" (Huberman, 2017; Williams, 2018; Wu, 2018). While smartphones are a central component of the attention economy, discussion of the relationship between smartphones and attention often centers on the impacts on behavior and task performance created by both the ever-present pull of smartphone content and interruptions from smartphone notifications.

Interruptions

Much like privacy, the study of interruptions is not confined to smartphone use. Monk et al. (2002) found interrupting a user as they were starting to work on a task had less negative impact on performance (i.e. switching cost) than interruptions in the middle or toward the completion of a task. Bailey & Konstan (2006) found hypothetical workers given periphery or non-essential tasks in-between primary tasks, out-performed participants who were interrupted with periphery tasks while in the middle of working on primary tasks. Additionally, participants in a condition that separated primary and periphery tasks also reported lower levels of annoyance and stress completing the tasks. While unsurprising on the surface, the authors' conclusion calling for "attention managers" (p. 705) to regulate interruptions and serve them at opportune moments (i.e., in-between tasks or after tasks have been completed) to help combat these effects have yet to be developed and implemented.

In commercial settings, Xia & Sudharshan (2002) found interruptions increase the time buyers spend on websites, but also lead to more difficulty in making purchasing decisions when they are incurred early in the buying experience and when users have abstract buying goals (e.g. electronics shopping instead of a specific item like a phone charger). Conversely, interruptions late in the buying process were generally ignored, presumably because users had made up their minds and wanted to proceed to purchasing and were less impactful. However, when interrupted/prompted by new product offerings just after making a purchase, users generally had favorable impressions and willingness to engage with the new information. Tsang et al. (2004) extending these findings showing users having negative attitudes toward SMS text advertising, but more favorable impressions when permission was obtained prior to sending and when ads were entertaining.

Interruptions from text messages are usually welcomed by users but have also been shown to have negative consequences. Rather than thinking of texting as a periphery task, users often view text messages as welcomed diversions (regardless of their impact on stress or task performance). Fischer et al.'s (2010) participants were unconcerned when texts interrupted them but were very concerned with the content they contained. Provided that participants found the content of text messages to be relevant, actionable and/or interesting, the negative effects of inopportune timing were mitigated. Missing from the study however were evaluations of users' feelings of stress, anxiety or being overwhelmed in relationship to the volume of texts or notifications they received.

Notifications

App notifications are the most common form of interruption from smartphones. "Push notifications" or system generated alerts from mobile apps may inform users of weather updates, breaking news, calendar reminders, new messages, social media interactions, or happenings in a video game (among many other informational notices). That is, the diversity of smartphone applications generates an equal diversity of alerts or notifications users may choose to address. The result of which is smartphone users fielding and managing a very high volume of notifications each day with relatively few technological options/settings for how to control when these notifications are received (Shirazi et al., 2014).

Much like the effect of interruptions discussed to this point, the multitude and randomness of mobile notifications have adverse effects on users' abilities to focus on and perform tasks (while also having perceived benefits). Stothart et al. (2015) showed even when users did not check their phones, merely hearing or seeing a notification come through led to significant task performance declines in comparison to control groups which did not have access to their smartphones while completing the same tasks. Although untested, the authors speculated participants were thinking about what the notifications could be alerting them to or having "task-irrelevant thoughts" which distracted and hurt their performance. Somewhat counterintuitively, research has also shown users respond more quickly to notifications when they are engaged in complex or cognitively challenging tasks (Mehrotra et al., 2016). Investigators hypothesized this may be a result of users being more aware or hyper-engaged when working on demanding tasks, which then makes them more attuned to noticing notifications. However, as previously shown, noticing these interruptions also leads to lower performance on tasks that are inherently more difficult to complete, thus creating something of a vicious cycle.

In addition to lower performance levels, Pielot et al. (2014) observed notifications leading to higher levels of perceived stress, annoyance and disruption. The type of notification received was shown to have big impacts on it being received positively or negatively. Negative perceptions were most associated with notifications from email and social media applications. Notifications from messaging apps like SMS text and WhatsApp were perceived as increasing feelings of connectedness and utility, showing again the contextual nature of users' relationships with their smartphones.

Summarizing, smartphone users are frequently interrupted by mobile notifications which are associated with both positive and negative emotions. Distraction from notifications has been shown to reduce attention levels (Zheng et al., 2014) and lower task performance, but may also increase perceptions of connectedness and productivity (M. A. Mazmanian et al., 2005). Ultimately, smartphone users are highly aware and responsive to notifications and may prioritize smartphone notifications above other tasks requiring their attention (Mehrotra, Pejovic, et al., 2016; Sahami Shirazi et al., 2014). By attending to their smartphones and mobile notifications frequently, users increasingly occupy a hybrid space in which they are present not only to individuals and features in their physical environments, but also to others and information in their digital networks and applications. That is, they are living in environments embedded with computing technologies which divide and blend their attention between cyber and physical space.

Managing Presence

The need for smartphone users to allocate attention between their physical surroundings and computing technology requires multitasking. Unfortunately, humans are not terribly good at multitasking, especially in situations where accuracy is important (Adler & Benbunan-Fich, 2012; Carrier et al., 2015; Hembrooke & Gay, 2003; Jeong & Hwang, 2016). When Oulasvirta et al. (2005) asked participants to multitask between various objectives in their physical environment while also navigating a web browser on a smartphone, they observed attention spans for both tasks dropping to as little as four seconds. In other words, common smartphone activities require a great deal of attention that is not easily shared with other tasks (i.e., highly distracting).

The smartphone's demand for attention has effects on both smartphone users and those in physical proximity to the user. It is the latter group that Tolmie et al. (2008) focused on in their ethnography of playing a SMS text based board game. While contextual norms varied, the overriding finding was that mobile notifications have significant impacts on family, friends and coworkers who are present with smartphone users when they receive notifications. They point out that neither party knows the nature of a generic tone or vibration when notifications arrive, but users' attention goes to their smartphone when notifications are investigated. This is generally interpreted, as smartphone users ignoring others physically present when they attend to their smartphones and this lack of attention to others generates attendant consequences to their relationships. While the authors propose the invention of notifications that can communicate to users and others physically present about their urgency and relevance (e.g., some technological equivalent of an assistant interrupting a business meeting with, "I am sorry Ms. Smith but your mother is on the line and says it is urgent"), these kinds of notifications have yet to be developed.

Smartphone users themselves are impacted by living in an "always on" and thereby "always accessible" state to others and technology. In a series of publications, M. Mazmanian et al. (2013) and M. A. Mazmanian et al. (2005) develop the term "autonomy paradox" to describe the effect smartphones are having on work behaviors in several client service industries. Through in-depth interviews with investment bankers, venture capitalists and lawyers, they make an argument that the perceived benefits of smartphones (i.e., convenient access to email, contacts, contracts, etc.) lead to negative outcomes including increased stress, less work/life balance and weakened familial relationships. Summarizing the paradox, the freedom to work anywhere and at any time has gradually led to the expectation (at least in the industries observed) that individuals should always be available to work. Importantly, professionals in these industries were seemingly unaware (or in denial) of any negative effects and attributed negative impacts on sleep, stress, family, etc. to being the costs of success rather than to smartphones negatively impacting their attention and time management capabilities. There is evidence to suggest these effects may not be contained to specific industries, but happening to society in general. Thomée et al.'s (2011) survey of young adults in Sweden showed high correlations between mobile phone use and various mental health concerns including depression, increased stress and sleep disturbance.

The expectation for individuals to always be available is not slowing down. Building on her earlier work, M. Mazmanian and Erickson (2014) argue that availability itself is becoming one of the key differentiating factors for many firms. Businesses are increasingly competing on who can provide clients with the most comprehensive access to their services 24 hours a day, seven days a week, 365 days a year. While theoretically obtainable in the short term, the authors point out how finite resources for human capital and attention make this is an unsustainable arms race. At some point, industry will run out of qualified professionals to hire and individuals can only provide constant access to their attention for so long before a breaking point is reached.

To combat these problems, M. Mazmanian and Erickson (2014) propose team approaches developed by Perlow (2012) and Perlow & Porter (2009) that simulate uninterrupted access to individuals or structures and allow individuals to have designated time off periods through team support systems that may address client needs with a certain degree of anonymity or ambiguity about whom is actually responding. Aoki & Woodruff (2005) also proposed ambiguity systems to help manage the accessibility issue or autonomy paradox. They suggested the use of "contact leases" which could terminate after a specified time or for a range of stipulated reasons in order to better manage which individuals/clients would be able to contact a phone at any given time. In all cases, these studies show the increasing demand for users' attention, time and autonomy stemming from the increased use and reliance upon smartphone technology.

Stepping back, privacy and the right to privacy are intimately related to notions of public and private individuals and public and private spaces. As the review to this point has illustrated, electronic media are complicating formerly held definitions and understandings of what it means to be "present" in pervasive computing environments as well as the distinction between what is meant by public and private individuals. Are smartphones serving to collapse society back into tribal villages? How do individuals maintain backstage (rehearsal) spaces in environments increasingly dominated by surveillance technologies? Are we all public figures in the era of smartphones and social media? What happens to autonomy in the face of big data operations that may know our wants, needs, emotions, locations, and health indicators before we do? What are the consequences to family, friends and physical intimacy within environments increasingly blurred by physical and digital presence? With the growing pervasiveness of computing technology and its processing capabilities, is anyone able to comprehend, much less deliver "informed consent" about how their information is collected and used? When our attention is tuned to interruptions, alerts, and reminders can we really be "present" anywhere in cyber or physical space? That is, how can we explore the myriad dimensions of the question:

RQ: How are millennials managing privacy, presence, and publicity in the face of digital

surveillance and "always on" connectivity?

A method for answering this question is presented in the following chapter. It builds on the theory of media effects first developed by McLuhan by investigating how his prediction of electronic media acting as "implosive" forces on privacy, presence, and place (i.e., public/private identities, accessibility to others, expectations for availability and the ability to focus or allocate attention) align with the perceptions of young adopters of modern society's most dominant media technology, the smartphone. The chapter begins with a discussion of surveys used to measure these perceptions and concludes with a discussion of how survey responses were screened and analyzed.

Chapter 4: Methodology

Data from two surveys will be reported in the chapter below. The researcher conducted the first survey in 2015, however he chose to take a leave of absence before completing its analysis. The study resumed in 2020 at which point the committee overseeing the research advised collecting current data to supplement the 2015 findings. The result of which was a second survey being administered in 2020. The 2015 survey was used as the basis for the 2020 instrument, however there were some changes included to reflect findings in the 2015 data and to incorporate questions raised in the literature published between 2015 and 2020. The 2020 survey was also conducted during the Covid-19 pandemic and needed to gather responses digitally as opposed to the in-person method used for the 2015 survey. Overall, the surveys were made as consistent as possible, but differences will be noted as needed in the following chapter.

The chapter begins with a brief description of each survey and how it was designed. Next, the process by which key concepts were operationalized or turned into variables is discussed. To streamline this section, the 2015 survey is always discussed first. Then, any changes made to the 2020 survey are presented immediately after. Once the method of the surveys development is concluded, there is a brief discussion of a pilot study conducted to troubleshoot comprehension issues, omissions, and completion time for participants in the 2015 sample. Samples for both surveys are then examined before moving to a discussion of data screening and analysis.

Ultimately, survey responses were used to generate linear regression models that identified relationships between smartphone use and operationalized concepts of privacy, presence, and publicity. As discussed, all three of these concepts are somewhat vague or contested. To improve the reliability and understanding of the results, principal component analyses were used to identify latent, or factor variables suggested by the responses. For example, correlations observed between responses to questions designed to measure "privacy" were used to parcel out the multiple dimensions of what privacy meant to participants. Once these dimensions or factors were identified, they were used as the variables entered in the regression models rather than responses to individual questions or aggregates of questions that combined multiple meanings/dimensions of the terms. A discussion of this screening process is given before the chapter concludes with a presentation of the assumptions, capabilities, and limitations of multiple regression modeling.

Survey Development

The 2015 survey consisted of 36 questions (Appendix A). Most of its questions were adapted from studies conducted by The Pew Research Internet Project. In particular, the work of Boyles et al. (2012), Madden (2014) and Rainie et al. (2013) were used to measure respondents' perceptions of privacy and publicity. A. Smith's (2012) questions were adapted for measuring perceptions of presence (i.e., attention, notification stress, accessibility, and access to others).

Responses were gathered using four-point Likert scales and Yes/No prompts (public identity was also gauged by the total number of social media sites respondents contributed to in an average month with a max score of seven). Four-point scales and "yes/no" questions were chosen to force respondents to indicate their agreement or disagreement with statements. That is, the instrument was designed to elucidate responses that indicated participants overall tendency toward being concerned about privacy or not, perceiving their smartphone as distracting or not, and maintaining a public identity or not without providing a neutral option. The removal of the neutral option is a contested decision in the survey design literature. Some research suggests that including a neutral option decreases extreme responses (Weijters et al., 2010) and increases response reliability (Lozano et al., 2008). On the other hand, research has shown respondents often pick the neutral option to avoid doing the cognitive work necessary to express their actual option (Krosnick et al., 2002), and that people often gravitate to the neutral option in order to avoid confronting their conflicting feelings about an issue (Bishop, 1987). Questions that require individuals to think about multifaceted elements of their lives and form an opinion have also been shown to generate higher levels of neutral response as a means for "satisficing" (Johns, 2005; Nowlis et al., 2002). Given previous observations of paradoxical relationships existing between privacy concerns and behaviors as well as between perceptions of productivity and decreased autonomy, the removal of the neutral option was justified by a desire to avoid respondents using it as a cognitive shortcut. Furthermore, the removal of the neutral option created an opportunity for statistical analyses that prior to analysis were considered as potential means for reporting results.

Smartphone use levels were measured by asking respondents to indicate how many minutes they spent using their smartphone during an average day for a variety of common activities that helped respondents think through their response. In total, the survey was designed to support a model that would use perceptions of privacy, presence, and publicity to predict smartphone use levels. Smartphone use was collected in a fashion to maximize accuracy and response variability. Perceptions of privacy, presence and publicity were measured using bounded responses that required respondents to indicate their level of agreement or disagreement with statements without a neutral option.

The 2020 survey began with the same questions being asked in the same order as the 2015 instrument. That said, it did omit questions which had little or no variance in responses from the 2015 participants. The 2020 survey also added several questions to address new literature and areas of inquiry prompted for by analysis of the 2015 responses. Several questions were created to better measure understandings of privacy permission and policies (Reidenberg et al., 2015). Questions asking about privacy agency and choice were inspired by Acquisti et al. (2017). Privacy apathy and futility perceptions suggested by Auxier et al. (2019) were also probed by 2020 survey. In addition to new questions about privacy, perceptions about smartphones impacting attention, productivity, and stress were added to address findings from the 2015 sample and the work of M. Mazmanian et al. (2013), M. Mazmanian & Erickson (2014), Perlow (2012), Pielot et al. (2014) and Stothart et al. (2015). Finally, several identity performance measures were added to probe deeper into perceptions about social media use and it is potential relationship with public identity perceptions. In total, the 2020 survey included 80 questions (Appendix B).

Both surveys were reviewed and approved by the researcher's university internal review board prior to administration. To increase responses to the 2020 digital survey, recruitment materials incentivized participation by entering respondents in a drawing for five $20 Amazon gift cards. After completing the drawing and rewarding participants for their participation, all respondents' data was anonymized by deleting their personally identifiable information from the dataset.

Hypothesized Relationships

Privacy concern or worry about not being able to control the collection and use of one's personal information was repeatedly observed by studies presented in chapter three. However, users were also observed to discount their concern when downloading applications and engaging in risky online sharing behaviors (privacy paradox). Given the "implosive" effects of electronic media theorized by McLuhan, privacy concern, operationalized as the level with which individuals worry about their personal information online, was hypothesized to have an inverse relationship to levels of smartphone utilization. Or that lower levels of privacy concern would be associated with higher levels of smartphone use. Although consistent with McLuhan's theory that electronic media amplify perceptions of being part of a collective global village rather than those of being a private individual, this hypothesis is at odds with empirical observations of the privacy paradox (i.e., where individuals engage in risky behaviors despite expressing high levels of privacy concern). As such the hypothesis here is uncertain as historical theory and more recent observations are at odds with one another. Regardless, privacy is an important variable to include when studying the effects of smartphones and is included in this study's models to test both McLuhan's theory and to compare with recent empirical observations.

H1: Privacy concern values will be negatively associated with total smartphone use.

In relation to individuals' smartphone usage, presence was operationalized in this study as the degree to which individuals felt their smartphone distracted their attention or made it difficult for them to focus on primary tasks. Additional presence measures asked if users felt smartphones made them more productive or enhanced their relationships with others and information. Given findings about the autonomy paradox and the effect of interruptions on focus, it was hypothesized that distraction would be positively related to smartphone engagement levels. That is, increased smartphone use would be associated with increased levels of distraction. With the framing of McLuhan's "global village," the researcher tested whether increased engagement with smartphones made it more difficult to separate from the digital networks/communities with which one is connected. H2: Distraction values will be positively associated with total smartphone use.

Based on previous research showing users often believe their personal information is being shared beyond their ability to control it, a hypothesis that users would report increased perceptions of themselves as maintaining public identities was generated. Combining McLuhan's (1964) theorizing with Cuff's (2003) notion of the "cyburg" results in the prediction that one's visibility and accessibility to others will be positively related to their level of smartphone utilization. To borrow McLuhan's imagery again, there are few places in a global village where one cannot be observed by the other village members. In this fashion, the author expected to find individuals perceiving themselves as having a public identity if they regularly utilized their smartphones. To further contextualize findings, responses were gathered to measure levels of identity performance in both digital and physical environments.

H3: Public identity values will be positively associated with total smartphone use.

Respondents' total level of smartphone use was chosen as the outcome variable of interest in this study. Participants were asked to provide the average number of minutes they spent using their smartphone each day. In order to aid participants' thinking about their response, the survey prompted them to indicate the minutes they spent engaging in an exhaustive list of potential activities (e.g. texting, emailing, social media, gamming, etc.). Each participant's responses were then summed to create a total smartphone use score to be regressed on the operationalized variables presented above to test the hypothesized relationships. There was also a question that asked participants to indicate their gender. There is evidence that suggests gender differences may impact smartphone use, privacy concern levels and engagement with social media platforms (Brell et al., 2016; Hoy & Milne, 2010). Consequently, these differences could impact the variables of interest and were controlled for in the study. The multiple regression section at the end of the chapter discusses how these relationships were evaluated in greater detail.

As with any survey, participant responses were influenced by their personal histories. Although the data analysis attempts to control for differences in gender, it is not possible to control for every individuals' life experience. Variables were confined to concepts and constructs suggested by McLuhan's theorizing about electronic media (specifically, smartphones). Based on this theory and related contemporary work, privacy, publicity and attention were expected to be significant predictors of smartphone use. However, It is never possible to be certain that one has correctly specified a statistical model (i.e. it contains all relevant predictors and does not include any irrelevant predictors | Warner, 2008). Indeed, testing the validity of media effects theory as a predictor of smartphone behaviors was among the desired aims of this work.

Pilot Study

Prior to collecting sample responses for the 2015 survey, two focus groups were conducted to refine the wording of survey questions, collect feedback about missing or incomplete items, enhance the survey design and estimate completion time. Most participants were able to complete the survey in five minutes or less and found questions to be clearly stated and concise. Several social media sites were added because participants indicated they used applications not prompted for in the initial design. Several questions were also revised to reflect student vocabularies more accurately. Terms like "jobs" and "work" were replaced by "class" and "school." A publicity question asking about how students search for information about others was revised to include the phrase, "After I meet someone new…" to better hone responses to reflect thinking about others in their social/professional networks.

Data screening from the pilot study did not show significant relationships between smartphone use level and the predictor variables. The researcher also believed that respondents were underestimating how much time they spent using their smartphone each day. The believed underestimation of smartphone use was addressed by adding the common smartphone activity prompts mentioned above (i.e., by asking respondents to indicate how much time they spent texting, calling, playing games, etc. instead of a simple estimate of how much time they spent using their smartphone each day). Given the number of focus group participants (16), it was expected that an increased N size for the sample along with the additional smartphone use activity prompts would generate greater statistical power in the final analyses.

Survey Samples

The 2015 survey data was collected in-person from a sample of 102 undergraduate students enrolled in a single media communications class. Although a convenience sample, the study of undergraduate students in media provides several unique insights that would not be possible using a more representative sample of the population. Primarily, most undergraduate students are between the ages of 18 and 22-years-old. At the time of the 2015 collection, a majority of the students were likely born between the years of 1993 and 1997. This population is interesting because they are the oldest (and thereby most experienced) individuals alive who have had access to the World Wide Web (WWW) for their entire lives (Hanson, 2011, pp.353-354). Having grown-up with the WWW, undergraduates were assumed to lack, or at least to have the least bias for, the effects of older media forms like print, radio, and television. Media students (all other things being equal) were assumed to be more knowledgeable about their smartphones, advertising methods and media regulations. Being familiar with media production techniques, issues of media literary, and media industry practices should give these students the ability make more informed and deliberate decisions about media consumption/production behaviors than the public at large. Other things held constant, sampling more savvy users was expected to measure more contemplated choices by the participants rather than a reliance on the default settings of smartphone applications, popular practices and/or uninformed interpretations of risks and benefits. Furthermore, this population was viewed as being predictive of the attitudes that will drive social and political policy in the coming decades. Given their lifelong exposure to the WWW and mobile communication technologies, this group is closer to being representative of the next generation's perceptions of the world. Understanding this population's views is essential for creating social, political, and educational policy in the era of pervasive computing technologies and for predicting the issues and debates that will arise from spaces and societies increasingly characterized by digital connectivity and surveillance.

For the 2020 sample, students were drawn from the same undergraduate course used for the 2015 sample. This was done to make comparisons as appropriate as possible and to maintain the advantages gained by sampling savvy media users. Unfortunately, sampling in 2020 was conducted in the middle of the Covid-19 pandemic and needed to be done electronically rather than in-person. It is generally understood that response rates go down significantly when surveys are conducted via email verses in-person (Daikeler et al., 2020). Indeed, even though the researcher received help from the sampled class's professor recruiting participants, only 35 of 197 students responded (18%). While disappointing from the standpoint of generating statistical power, it was nonetheless greater than the floor of 30 identified by Roscoe (1975) and supported for internet survey research by Hill (1998). Furthermore, since the sample was drawn from the same class at the same university, overall similarities in responses would support the understanding of both samples as being representative of one another (and thereby as representative of undergraduate students studying media). The 2020 respondents were presumed to be about five years younger on average than those in the 2015 sample and were assumed to be highly competent with digital communication technologies. Samples for both surveys skewed to the male perspective. The 2015 responses consisted of 52 males (57%) and 39 females (43%). The 2020 responses consisted of 23 males (66%) and 12 females (34%). In both cases, males were overrepresented, and results need to be interpreted with this limitation in mind.

Data Screening

Responses were received from 102 participants in the 2015 data collection. Of these, four reported not owning a smartphone and two did not indicate if they owned a smartphone or not. Responses from these six participants were dropped as the low number of individuals not owning a smartphone prohibited meaningful comparisons between those who did and did not own smartphones. Revising the smartphone use measure to prompt respondents about different activities did increase the amount of time participants reported using their smartphones compared to the levels seen in the pilot study. However, five participants did not provide any response to the more detailed (and thus more cognitively demanding) prompts. Responses from these individuals were also dropped from the analyses which resulted in usable data from a total of 91 respondents.

Since the 35 responses gathered by the 2020 sample were collected electronically, the researcher was able to force respondents to provide valid responses to all questions. Consequently all 35 participants provided answers to all the survey questions. While beneficial in the sense of generating 100% compliance with the survey instrument, the lower response rate (only 35 of a class with an enrollment of 197 or 18%) is clearly a drawback of electronic survey collections.

Data screening for implicit variables found in each collection (i.e., the 2015 and 2020 surveys) will be presented separately in the following sections. Variables identified in the 2015 responses will always be presented first and followed by their 2020 counterparts. As the 2020 survey expanded the range and types of questions participants were asked, it generated additional variables that could be modeled in analyses. That said, the larger sample size of the 2015 survey generated greater statistical power for the 2015 findings. In all cases, the researcher carefully analyzed the results to generate variable names and groupings that had intuitive meanings and were consistent with the assumptions of principal component analyses.

2015 Smartphone Use Variable Screening

The smartphone use prompts may have caused several participants in the 2015 survey to overestimate the total time they spent using their smartphone on an average day. From the 91 completed surveys, the mean amount of time reported using a smartphone was 407 minutes or six hours and 47 minutes each day (median time 305 minutes and mode 240 minutes). While it was not surprising to see high levels of use, sample responses deviated from a normal distribution due to the presence of several outliers (including one user reporting a total of 3,650 minutes or 60 hours of use per day). This was seen both through visual examination of histograms and high index values for skewness and kurtosis. All these screenings indicated a strong positive skew and a sharp peak in the response distribution. To identify outlier values, standardized scores were computed for each respondent on the smartphone use variable. Z scores above 3.2 (all outliers were positively skewed) were removed until all responses were within normal limits. After the removal of outliers, all Z scores for smartphone use were between -1.54 and 2.46. Visual inspection of the response histogram was also qualitatively evaluated to be approximately normal. The effect of this culling was the removal of five participants' responses that reported using their smartphone for more than 15 hours (900 minutes) per day. Although this removed values that were technically feasible (i.e., less than 24 hours), the resulting maximum value of 766 minutes or roughly 13 hours seemed a more realistic estimate that still preserved representative variance of the sample. At the conclusion of screening and outlier removal, 86 responses with smartphone use or "Y" values suitable for linear modeling were retained.

2020 Smartphone Use Screening

The 2020 survey prompted 11 smartphone activities thought to be exhaustive of how individuals may use their smartphones along with one catchall prompt ("other") to capture respondents' total amount of smartphone use on a typical day. The mean level of smartphone use for the 35 respondents was 493 minutes or about 8 hours and 13 minutes per day. Z scores for each response ranged from -1.3 to 2.8 and were within normal limits, however visual inspection of the response histogram and evaluation of the index values for skewness indicated a slight positive skew. The removal of a single outlier reporting 1172 minutes of daily use (19 hours and 30 minutes per day) generated a set of responses that were evaluated to be normally distributed. This created 34 smartphone use values with a maximum of 900 minutes (15 hours) and a minimum of 180 minutes (3 hours) that were suitable for linear modeling.

Predictor Variable Computations

Concepts like privacy, presence, and publicity are inherently vague and contested terms. As such, when creating a survey to measure these concepts it was assumed that questions would measure multiple dimensions of these terms. Factor analysis is a method for reducing large number of variables (or responses in the present case) to a smaller set of common factors that can be used as score variables. For example, it provides a method for grouping multiple questions designed to measure "privacy" into intuitive categories that parcel out its multiple dimensions. Once these dimensions are identified and evaluated, they can be used as variables in statistical modeling. The identification of dimensions, also known as factors or latent variables, is done by evaluating response correlations between different questions. Principal component analysis (PCA) is a common method of factor analysis and was used to identify variables used in this study. PCA is based on the general linear model and makes assumptions that the data is normally distributed, variables are linearly associated, and is not influenced by extreme outliers. The next several sections report how PCAs were used to generate variables for this study's analyses.

2015 Privacy Variables

Participants' online and mobile privacy beliefs were measured by responses to 12 questions[1] in the 2015 survey instrument. Examination of response histograms to privacy question responses showed some deviations from normal distributions, however the

____

deviations were not deemed prohibitive for conducting principal component analyses (PCA). Responses from the 91 total completed surveys were used in a PCA that identified factors or variables with eigenvalues greater than 1. Varimax rotation was requested to aid in the interpretation of variable categories. Nine of the privacy questions gathered responses on four-point Likert scales and three gathered responses using "yes/no" prompts.

The PCA identified four latent variable groupings or factors that collectively accounted for 66% of the total variance in privacy responses. All the groupings had intuitive meanings and were named by the researcher. "Secure application" perceptions were measured by five questions[2] and accounted for 29% of the variance in responses. Three questions[3] measured "privacy concern" and accounted for 16% of the variance in responses. Three different questions[4] measured "privacy protection behaviors" which accounted for 12% of the total variance and a single question measuring "anonymity online" accounted for 8% of the total variance. "Secure application" correlation loadings ranged from .56 to .80. "Privacy concern" loadings ranged from .67 to .80. Loadings for "privacy protection behaviors" ranged in absolute value from .56 to .89 and the single item "anonymity" loading was .96. Component correlation matrixes are found in Appendix C.

Factor values/scores were computed by summing responses to questions correlated with each latent variable category (using reverse coding in cases of negative loadings). For example, the secure platform variable is the summation of replies to five questions which used four-point Likert scales (i.e., maximum value of 20 and minimum value of five). Both privacy concern and protection behavior variable scores are summations of responses to three questions. Anonymous activity represented a response to a single item and was omitted from analysis due to low reliability and relying on a single response for scoring.

2020 Privacy Variables

The 2020 survey instrument measured participants perceptions and concerns about privacy using 31 questions[5]. The response frequencies to privacy measures were approximately normally distributed and deemed suitable for PCA. Responses from all 35 responses were used in a PCA that yielded nine latent variable structures and accounted for 78% of the total variance in responses. Varimax rotation was used to aid in the interpretation of correlation loadings and facilitate intuitive naming. Responses to six questions[6] measuring "perceived security" accounted for 19% of the variance (absolute value of loading correlations loadings ranged from .5 to .9) in privacy perceptions. Responses to five questions[7] measuring "desire for regulation" accounted for 16% of the total variance (loading correlations from .5 to .8). Responses to four questions[8] measuring "privacy apathy" accounted for 10% of the variance (loading correlations from .5 to .9). Responses to three questions[9] measuring "privacy competence" accounted for 7% of the variance (loading correlations from .6 to .9). Responses to four questions[10] measuring "lost control of privacy" accounted for 6% of the variance (loading correlations from .5 to .8). Responses to three questions[11] measuring "social sharing concern" accounted for 5% of the variance (loading correlations .6 to .8) and four questions[12] measuring "privacy worry" accounted for an additional 5% (loading correlations .6 to .7). "Government surveillance' and "anonymity" were each measured by one question and accounted for 5% and 4% of the variance, respectively. Mirroring the process used for the 2015 responses, factor scores were computed by summing response values to questions associated with each factor category and reverse coding in cases of negative correlation/loading values. Rotated factor matrixes are shown in Appendix D.

2015 Presence Variables

Participants' perceptions of presence and attention were measured by 13 survey questions in the 2015 survey instrument. All the responses were gathered using 4-point Likert scales with the exception of one measured with a "yes/no" response. Like the privacy measures, visual inspection of histograms was qualitatively judged as being acceptable for PCA. The PCA of responses to presence questions revealed four latent variables which accounted for 63% of the total variance observed in presence question

---

responses. Varimax rotation was used to make loading correlations more interpretable. After rotation, intuitive meanings for all identified factors were identified and then named by the researcher. "Accessibility to others" was measured by four questions[13] and explained 31% of the total variance in responses. "Access to others" was also measured by four questions[14] and accounted for 15% of response variance. Responses to three questions[15] measuring "attention/distraction" due to smartphones and accounted for 9% of the variance. Finally, "notification stress" was measured by two questions[16] which accounted for an additional 8% of the variance in responses. "Accessibility to others" loadings ranged from .57 to .82. Loadings for "access to others" ranged between .60 and .68. Correlations for questions about "attention/distraction" from smartphones ranged from .57 to .82. Finally, "notification stress" correlation loadings were .73 (for both questions). In the same fashion as privacy factors/variables, presence factor scores were computed by summing responses to questions correlated with each latent variable category and used in place of individual question responses in the models reported in the results section. None of the presence factors had negative loadings and therefore did not necessitate reverse coding prior to summing. Results for the 2015 PCA of presence are shown in Appendix E.

2020 Presence Variables

The 2020 survey used 31 questions[17] to measure presence perceptions. Response frequencies were approximately normal and suitable for PCA. Varimax rotation was used to make structure interpretation easier after which intuitive names for each variable were created by the researcher. The PCA identified nine latent variable structures that accounted for 79% of the total observed variance in presence perceptions. "Enhanced functionality" was measured by six questions[18] and accounted for 27% of response variance (loading correlations ranged from .5 to .9). "Lost focus" was measured by seven questions[19] and accounted for 13% of response variance (loading correlations ranged .5 to .8). "Habitual behavior" was measured by responses to three questions[20] and accounted for 9% of the variance (correlation loadings .6 to .9). "Constant connection" was measured by responses to three questions[21] that accounted for 7% of presence response variance (loading correlations from .6 to .7). "Information overload" was measured by three questions[22] that accounted for 6% of the variance (loading correlations from .5 to .9). "Obligation stress" was measured by two questions[23] that accounted for 5% of the total response variance (loading correlations .6 to .9) and "Organized productivity" was measured by three questions[24] that also accounted for 5% of the variance in responses (loading correlations .4 to .9). "Disconnected" was measured by responses to three questions[25] that explained an additional 4% of the variance in responses (loading correlations .5 to .8). Finally, "friendship" was measured by a single item that explained 3% of the variance. Factor scores for each latent variable were generated by summing responses to all questions associated with each structure and reverse coding was performed prior to summing in cases of negative correlation/loading values. As "friendship" was only associated with a single question, it was not generated as a latent variable for analysis. Appendix F shows the rotated component matrix for the 2020 presence variables.

2015 Publicity Variables

The 2015 survey used seven questions to measure participants' perceptions of publicity or public identity[26]. Except for the total number of social media accounts (which could range from zero to seven), all other responses were gathered using 4-point Likert scales. Response frequencies were found to meet the assumptions of PCA (normally distributed). A PCA of the responses that utilized varimax rotation of the loading correlations identified three factors that explained 64% of the observed variance in publicity perceptions. All three factors were interpreted and named by the researcher. "Publicity behaviors" were measured by three questions[27] and accounted for 27% of the total variance. "Performance behaviors" were measured by three questions[28] and

____

accounted for 22% of the response variance. Finally, a single item measured "invisibility" and accounted for 15% of the total variance in publicity perceptions. "Publicity behaviors" had loading correlations between .61 and .80. Loadings on "Performance behaviors" ranged from .42 to .83 and the single item "invisibility" loaded at .90. Publicity and performance factor scores were both computed by summing responses to the three questions which made up their structures (i.e. using the same method as all predictor variables discussed above). The invisibility factor was omitted from analysis due to low reliability as it relied on single response for scoring. Results from the 2015 publicity PCA are provided in Appendix G.

2020 Publicity Variables

The 2020 survey asked participants to respond to 15 questions designed to measure publicity perceptions. Except for the total number of social media accounts, all questions measured responses using four-point Likert scales. Analysis of the response frequencies showed they did not violate the assumptions of PCA. A PCA of responses to all 15 questions revealed five latent variable structures that accounted for 66% of the total variance in publicity perceptions. Loading correlations were rotated using the varimax method and the five factor structures were named by the researcher. "Social media identity" was measured by responses to four questions[29] and accounted for 25% of the total variance of publicity perceptions (loading correlations ranging from .6 to .8).

"Public figure" was measured by responses to four questions[30] that accounted for 13% of the variance (loading correlations .5 to .8). "Social media stress" was measured by responses to two questions[31] and accounted for 10% of the variance (loading correlations approximately .8). "Selective social-presentation" was measured by three questions[32] that accounted for 9% of the variance (loading correlations .5 to .7) and "authentic sharing" was measured by two questions[33] whose responses accounted for 8% of the total variance (loading correlations approximately. Factor scores for all five factors were computing by summing response scores. Appendix H shows the rotated component matrix for 2020 publicity variables.

Data Interpretation and Modeling

The creation of factor variables discussed above was done for two primary purposes. Given the number of questions included in both surveys, and particularly the 2020 survey, the creation of factor variables allowed for correlations to computed with a much lower likelihood of generating type I errors. For example, correlations between all 80 questions included in the 2020 survey would be expected to generate $\frac{(80+79)}{2} \times .05 \approx$ 4 erroneous significant correlations due to type I error. By limiting the number of correlations being computed to those between 23 factors, only one erroneous relationship was expected.

The second reason for creating factor variables was a desire to represent concepts like privacy concern, distraction from smartphones, and public identity perceptions with valid constructs. As noted previously, all these concepts may have multiple meanings. As both surveys asked questions prompted for by a variety of studies in the literature review, the factor analyses were a method for narrowing the interpretation of results into specific categories generated by correlations of the participants' responses. For example, the factor analysis of privacy questions in 2020 operationalizes the concept of "privacy worry" into a value measured by aggregating responses to four specific questions that the data suggest are representative of a single factor or variable in the minds of participants. It is debatable whether these questions are exhaustive of every conceivable privacy worry that results from smartphone use, but the factor provides a definition of the term that can be used for statistical modeling. To test the hypothesis in the present study, these variables were modeled using multiple regression analysis.

Multiple Regression Models

Multiple regression analysis was chosen to create a model for understanding the relationship between privacy, presence and publicity with smartphone use levels. Formally, smartphone use level was regressed on four predictor variables: privacy concern, attention/distraction perceptions, publicity perceptions and gender in the study's central model. Multiple regression modeling was chosen because it provides measures for how well the entire model and each individual predictor variable predict the variance observed in the outcome variable (i.e., smartphone use) in the presence of the other independent variables. In other words, multiple regression is helpful for answering two related questions: (1) taken together, does the combination of privacy concern levels, attention/distraction perceptions, publicity behavior and gender predict a significant amount of variation in smartphone use levels and (2) how much variance in smartphone use is uniquely predicted by privacy concern, attention/distraction levels, publicity behaviors and gender? Slope coefficients, "b" in the model below, reflect how strongly each predictor variable predicts smartphone use while controlling for the other predictor variables in the model. Alternatively, "b" indicates how much change in smartphone use/behavior is predicted for every one-unit change in its respective variable (privacy concern, attention/distraction, publicity, and gender). Note that the intercept term in the model (b_0) is not meaningful because zero is not a possible score for the predictor variables (they represent sums of Likert scale responses, not scaled measurements).

$$Y_{(smartphone\ use)} = b_0 + b_{1(privacy)} + b_{2(attention/distraction)} + b_{3(publicity)} + b_{4(gender)}$$

Privacy concern is the first predictor variable evaluated by the model. Given the "implosive" effects of electronic media theorized by McLuhan, it was expected that privacy concern, operationalized here as the level at which individuals worried about their personal information online, would be negatively related to levels of smartphone utilization. In other words, the hypothesis that increased levels of privacy concern would predict decreased levels of smartphone utilization or use behaviors was tested.

Presence was operationalized in the model as a measure of attention/distraction. It is therefore understood differently than in many traditional definitions of the term. Typically, it is meant as description of physical placement or location. However, there are a variety of reasons one's mind may not be fully "present" within their immediate surroundings (daydreaming, reading a book, sending a text message, etc.). Today, there is also the potential for individuals to be engaged with digital or digital/physical hybrid environments generated by video games (e.g. Pokèmon Go), video messaging applications (e.g. Skype), and mobile applications (e.g. Geocaching) that further complicate understandings of presence as physical proximity.

In relation to individuals' smartphone usage, presence was treated as the degree to which smartphone were perceived as being distracting or making it difficult to focus. It was predicted that presence/distraction perceptions would be positively related to smartphone engagement levels. If we are moving to some version of McLuhan's "global village" through our adoption of electronic media, we should expect to find it more and more difficult to separate ourselves from the digital environments that make our global networks possible.

The final conceptualized predictor variable was publicity perceptions or public identity. Combining McLuhan's theorizing with Cuff's (2003) notion of the "cyburg" resulted in the prediction that one's visibility and public identity be positively related to their level of smartphone utilization. Using McLuhan's imagery, there are few places in a global village where one cannot be observed by the other village members. In this fashion, it is expected to find individuals perceiving themselves to have elements of a public/celebrity identity if they regularly utilize their smartphones.

In addition to the main model presented above, several additional multiple regression models were tested after considering new research and theoretical considerations that happened between the 2015 and 2020 data collections. At a general level, heavy amounts of smartphone use were assumed to be ubiquitous by the time of the 2020 collection. Given the ubiquity, the researcher wanted to explore smartphone use levels as a predictor rather than an outcome variable for analysis. As such, the latent variables for both the 2015 and 2020 were analyzed for suitability as outcome rather than predictor variables. Analysis of each variable's index value for skewness and kurtosis along with visual inspection of each variable's frequency distribution was performed. Although some small deviations from normal were observed in the 2015 (protection behavior and distraction scores) and 2020 (focus, digital existence, and social media identity scores) variables, the deviations were not deemed strong enough to prohibit linear regression analyses.

Significance testing (using alpha level .05) was used for all regression models. The overall model or omnibus analysis is a test of the hypothesis that the predictor variables (in combination) are unrelated to the outcome variable. When the omnibus test is significant, it indicates there is a predictive relationship between the predictor variables and outcome variable. The combined proportion of variance predicted by the predictor variables in the outcome variable is estimated by the term R^2. The unique variance predicted by each individual predictor variable in the outcome variable is estimated by semi-partial or part correlations as all variables were entered simultaneously in all models (i.e., all predictors are controlled for in the models). Part correlations were also tested at a .05 alpha level for significant difference from zero. Regression slopes (*b*) indicate the direction the relationship with the outcome variable in each model (i.e., positive or negative). SPSS was used to generate all test statistics and resulting *p*-values.

As previously discussed, smartphone use responses were culled to remove outliers and create an approximately normal distribution. Scatter plots were created for each predictor variable in relation to smartphone use to screen for linear relationships and homogeneity of variance for all predictors at each level of smartphone use. Scores were qualitatively interpreted to be within acceptable limits for linear analysis. Variance inflation factors (VIF) were generated by SPSS to screen for multicollinearity in all regression models. All were above one and below ten, which are the thresholds Ethington (2009) identified as the screening values for the presence of multicollinearity. In sum, the basic assumptions for linear regression modeling were met by the responses analyzed.

While the assumptions for linear regression modeling were met, the study does have statistical power limitations due to sample size. Warner (2008) cites multiple sources that recommend sampling at least 108 individuals for multiple regression models with four predictor variables where at least a medium effect size is expected. The study's samples of 86 and 34 responses are thereby below recommended levels. That said, similar findings in both studies supports the validity of the study's findings. As with any study, the results presented in subsequent chapters need to be tested in future studies that will ideally utilize larger samples.

Method Summary

Surveys were created to measure variables related to media effects predicted by the theorizing of Marshall McLuhan. Given the prevalent nature of electronic media in the modern living environment, it was imperative to understand how engagement with new media technologies like smartphones were affecting perceptions of self and society.

To address this question, the results in the following chapters present descriptive statistics and regression models using 86 responses to a 36 question 2015 survey instrument and 34 responses to an 80 question 2020 survey instrument. Four categories of measurement were central to the survey design. Users' perceptions of their digital information privacy, ability to allocate and manage focus/attention and public identity. Generally, these perceptions were reported through responses to questions that used Likert scales and "yes/no" response types. Respondents' total level of smartphone use was measured by the total number of minutes they spent using their smartphone on an average day. Respondents' gender was also collected to control for known differences in female/male communication behaviors and perceptions. In all the reported models, predictor variables (except for gender) are factor/latent variable scores computed by summing responses to multiple questions or summed scale responses to the smartphone use activities. Gender was always coded zero for males and one for females. All descriptive and inferential statistics use an n size of 86 and 34 for the 2015 and 2020 results, respectively. In both cases, outliers on the smartphone use variable were removed to allow for multiple regression modeling and to identify relationships with the factor variables and smartphone use levels. The following two chapters will gradually report more granular views of the survey responses. Descriptive statistics from both surveys are presented in the following chapter before statistical models with inferential statistics are presented in chapter six.

Chapter 5: Descriptive Statistics from Survey Results

Results from the 2015 and 2020 surveys will be reported in the following two chapters. This chapter will focus on descriptive statistics and highlight similarities and differences between the two samples. As previously mentioned, longitudinal comparisons are limited by the small sample size in 2020. Since the 2020 sample met the minimum requirements identified by Hill (1998) for internet survey research, the findings are presented with an assumption that responses are representative of the same population measured in 2015, but a future study with a larger sample size is desirable for confirming the 2020 results. General trends and descriptive statistics are presented by topic in the following sections. The chapter begins with an overview of the amount of smartphone use reported by participants in both samples. Privacy findings are presented second. Presence perceptions are presented next and publicity observations conclude the chapter. In every section, a general discussion of the findings precedes a table showing frequencies for the 2015 and 2020 results side-by-side.

Descriptive Statistics for Smartphone Use Levels and Reported Behaviors

This section reports respondents' average levels of smartphone use overall and for specific smartphone functionalities. Responses from the 2015 sample are presented in isolation before moving to a comparison with the 2020 results. Several response prompts were alerted, omitted, and added between the 2015 and 2020 surveys in response to growth in technological capabilities and evolutions in use trends between the two data collections. Average use time in minutes is summarized for both surveys in Table 1 below.

Participants reported using their smartphones for an average of 326 minutes or about five hours and 30 minutes each day in 2015. Use levels showed considerable variance with a minimum value of slightly less than one hour (50 minutes) and a maximum value of nearly 13 hours (766 minutes). Respondents spent the most of their time using smartphones for text messaging in 2015. On average, participants reported texting for about an hour and 15 minutes every day. Listening to music was also very popular with an average of 70 minutes. Participants spent an average of 65 minutes each day using their smartphones for social media activities and close to 40 minutes surfing the web. Reading, talking, emailing, and gaming were far less popular uses. The mode use time for reading and gaming were both zero (reading mean ≈24 minutes | gaming mean ≈ 11 minutes). Respondents spend an average of 22 minutes making phone or video calls (calling into question the name smart*phone*) and 20 minutes emailing during an average day.

There was a notable increase in the reported time participants spent using their smartphones in 2020. The average total time spent using a smartphone in 2020 was almost two and a half hours more than what was observed in 2015. As the sample was much smaller in 2015, and additional use prompts were added to the 2020 instrument, this difference needs to be interpreted with caution and confirmed by future study. That said, 2020 respondents reported using their smartphones for an average of 473 minutes or 7 hours and 54 minutes every day. Like in 2015, a wide range of use levels were reported ranging from three hours to over 15 hours per day.

Respondents in 2020 reported not only spending more time on their smartphones but allocating this time differently between smartphone applications. Whereas social media was the third most common use in 2015, it was far and away the dominant smartphone activity reported in 2020. On average, participants spent over two hours engaging with social media through their smartphones each day. This was double the amount of time reported in 2015. Listening to music was still popular with an average time of one and a half hours daily and was followed closely by video streaming. Average time spent using messaging applications fell to around 30 minutes per day (perhaps being replaced by social media messaging) and web surfing/reading was also down slightly to an average of 30 minutes daily. The remaining activities were generally less popular. Shopping, health applications, gaming and "other" activities all had mode values of zero and median values of 10 minutes or less. Participants said they averaged about 23 minutes a day making calls (audio and video), 13 minutes a day in production activities (taking photos, video and/or audio) and about 10 minutes for utility functions (e.g., maps, weather, and calculator). Perhaps the most poignant finding regarding smartphone use came from a question added to the 2020 survey to understand the smartphone's consumption relationship to more traditional mediums. A full 86% of respondents agreed that they used their smartphone more than any other media device (e.g., TV, radio, desktop/tablet, Roku/Apple TV, print) with 54% expressing strong agreement with this sentiment. In total, participants in both surveys showed high levels of smartphone use which supports the argument for understanding how this relatively new media technology is impacting individuals' perceptions of self, their environment and society.

Table 1

Smartphone Use Levels and Response Frequencies in 2015 and 2020

Survey Question	2015	2020
Mean total time respondents spent using their smartphone each day.	326	473
On average, approximately how many minutes do you spend using your smartphone to send/receive text messages each day?	76	n/a
On average, approximately how many minutes do you spend using your smartphone to send/receive messages or email each day?	n/a	34
On average, approximately how many minutes do you spend using your smartphone to listen to music each day?	70	98
On average, approximately how many minutes do you spend using your smartphone to use social media apps each day?	65	134
On average, approximately how many minutes do you spend using your smartphone to surf the web each day?	38	n/a
On average, approximately how many minutes do you spend using your smartphone to read (e.g., use e-readers, news apps, etc.) each day?	24	n/a
On average, approximately how many minutes do you spend using your smartphone to read/web surf (e.g. blogs, websites, news articles/apps) each day?	n/a	30
On average, approximately how many minutes do you spend using your smartphone to make phone calls (or FaceTime/Skype) each day?	22	23
On average, approximately how many minutes do you spend using your smartphone to send/receive email each day?	20	n/a
On average, approximately how many minutes do you spend using your smartphone to play games each day?	11	24
On average, approximately how many minutes do you spend using your smartphone to watch streaming video (e.g., Netflix, YouTube, Hulu) each day?	n/a	94
On average, approximately how many minutes do you spend using your smartphone to take pictures, record video, and/or audio each day?	n/a	13
On average, approximately how many minutes do you spend using your smartphone to make quick calculations (e.g., maps, calculator, weather) each day?	n/a	10
On average, approximately how many minutes do you spend using your smartphone to shop or buy food each day?	n/a	6
On average, approximately how many minutes do you spend using your smartphone to do an activity not listed (i.e., other) each day?	n/a	4
On average, approximately how many minutes do you spend using your smartphone to evaluate your health (e.g., step counter or sleep monitoring) each day?	n/a	4

Note. Columns report participants' mean average time in minutes for each survey question. Outliers were removed from analysis. Results in 2015 ($n = 86$) and results in 2020 ($n = 34$) have different sample sizes.

Descriptive Statistics Related to Privacy Perceptions and Behaviors

This section focuses on survey findings related to privacy. A complete listing of privacy response frequencies is show in Table 1 below. Overall, participants in both the 2015 and 2020 samples expressed similar levels of privacy concern and privacy protection behaviors. The survey results from both years indicated that most respondents had privacy concerns in relation to using smartphones. Nonetheless, a substantial number of respondents did not share these concerns. When respondents were asked directly if they were worried about the safety of their personal information, about half expressed worry and half did not. Similar levels of concern were expressed about the government's monitoring of phone calls and internet communications. Large majorities also desired anonymity for at least some kinds of online activities.

Although privacy was a concern for slightly more than half the respondents, a greater percentage of respondents had taken active measures to protect their privacy, including when making decisions about installing apps to their smartphone. Nearly 70% of respondents in both samples said they chose not to install an app due to is permission requests. Most respondents also reported taking this a step further by uninstalling an app after learning it was collecting information they did not want to share.

When asked about specific applications on their smartphones, respondents' level of privacy concern varied. Most participants felt secure sharing personal information on applications like Skype and Facetime and most also perceived text messaging applications as being secure. Participants were split on the perceived security of email, with about half thinking it was secure and half feeling it was not. Respondents overwhelming agreed that social media apps were not secure applications for sharing private information. A mere 20% and 11% felt secure sharing private information on these applications in 2015 and 2020, respectively.

A notable difference between the 2015 and 2020 results was the proportion of respondents having the location tracking feature of the smartphones turned on or off. This finding is notable because location data is often perceived as being highly sensitive to users (Bilogrevic & Ortlieb, 2016). Whereas only 43% of respondents said they typically had location tracking turned on in 2015, this jumped to 96% in 2020. Importantly, the 2020 survey provided respondents with an option to indicate, "on only while using" whereas the 2015 survey did not (as this feature only became readily available in the years post 2015). Nonetheless, the sharp rise shows that many more respondents in 2020 were sharing at least some location data with applications than what was observed by the 2015 survey.

The 2020 survey also asked 19 additional questions that were not included in the 2015 instrument. These questions bring nuance and additional understandings about the levels of concern and protection behaviors observed in 2015. Primarily, it is notable that most respondents reported reading and understanding the permission requests of applications before downloading them to their smartphones. However, only one-third of respondents indicated they read permissions requests when updating their apps. This implies that once an application is downloaded to a device, sample respondents were much less likely to notice changes to its data collection practices.

Responses to the 2020 additions also made it clear that most respondents felt high levels of uncertainty and low levels of control over how their personal information was being utilized. Large majorities said they had little control over how governments and application developers were using their personal information. In fact, large majorities simply assumed that businesses and the government were tracking most of their personal information and have given up on reading applications' privacy policies. Yet, what most united respondents was their agreement that they really do not know how much information their smartphone shares with businesses about them. This finding was supported by large majorities finding privacy policies difficult to understand and desiring additional legal protections over how their smartphone data could be used by governments and businesses. Collectively, these findings indicate that confusion may be the most accurate term for describing participants' perceptions of privacy in the smartphone era.

Finally, although participants expressed low levels of control and understanding about how their private information was being harvested by the big data industry, they did express behaviors that controlled how information about them was shared with individuals in their social networks. Large majorities said they were very careful about what information they posted to social media and restricted who in their social networks would be able to see their posts. Thus, while individuals may be unaware or apathetic about how businesses and governments may be using their personal information, they were highly attuned to methods of controlling how their information was accessed by individuals in their social circles. This seems to indicate that if individuals were given the tools to control how their personal information could be accessed by businesses and governments, they would be inclined to do so, but in the current environment these tools are either unavailable or make many popular applications unusable.

Table 2

Privacy Questions and Response Frequencies in 2015 and 2020

Survey Question	2015	2020
I worry about how much information is available about me on the internet	57%	60%
People should have the ability to use the internet completely anonymously for certain kinds of online activities	83%	91%
I have decided not to install a smartphone application (app) because I found out I would have to share personal information in order to use it	68%	69%
I have uninstalled an app on my smartphone because I found out it was collecting information I didn't want to share	70%	60%
Typically, I have the location tracking feature on my smartphone turned on	43%	96%
I am concerned about the government's monitoring of phone calls and internet communications	60%	54%
I feel secure sharing private information via text messages	63%	54%
I feel secure sharing private information via email	47%	51%
I feel secure sharing private information via applications like Snapchat	34%	43%
I feel secure sharing private information via applications like Skype and FaceTime	67%	69%

Table 2 Continued

Survey Question	2015	2020
I feel secure sharing private information via social media applications (e.g. Facebook & Twitter)	20%	11%
I worry a great deal about the safety of my personal information online	57%	57%
I usually read the privacy policies of the websites on which I make purchases	n/a	31%
I usually read the permission requests of the apps I download to my smartphone	n/a	60%
I usually read the permission requests when updating apps on my smartphone	n/a	34%
I understand the privacy policies of the websites I frequently use	n/a	31%
I understand the permission requests for the apps on my smartphone	n/a	60%
I have little control over how my personal information is used by smartphone application developers	n/a	77%
I have little control over how the government uses my personal information	n/a	89%
My smartphone makes it harder to control who has access to my personal information	n/a	77%
I really do not know how much information my smartphone shares about me with businesses	n/a	94%
Smartphones make it impossible to have privacy	n/a	57%
I assume most of my personal information is tracked by the government	n/a	77%

Table 2 Continued

Survey Question	2015	2020
I assume most of my personal information is tracked by businesses	n/a	77%
I am very careful about what information I post to social media	n/a	77%
I usually restrict who can see my posts on social media	n/a	77%
There should be more legal protections over how smartphone data can be used by government agencies	n/a	94%
There should be more legal protections over how smartphone data can be used by businesses	n/a	94%
It would be impossible to thoroughly read all the permission statements I'm asked to agree to on my smartphone	n/a	74%
The language used in privacy policies is difficult to understand	n/a	89%
I have given up on reading applications' privacy policies	n/a	80%

Note. Percentages represent proportion of respondents that agreed with each survey question.

Descriptive Statistics Related to Presence Perceptions and Behaviors

The following section discusses responses to questions designed to measure perceptions of presence in relation to smartphone use. A complete listing of response frequencies described in this section is found in Table 2 below. As will be seen, responses to several of the questions included in the 2015 instrument showed little to no variance. As a result, only six of the 13 questions used in 2015 were carried over to the 2020 instrument. This section begins with a presentation of trends seen in the 2015 responses. Then, notable differences between 2015 and 2020 responses are identified. The section concludes by discussing responses to the 25 questions added to the 2020 survey that were not asked in 2015.

The 2015 responses strongly support an understanding of smartphones increasing users' access and accessibility to others and information. Virtually all respondents indicated they felt like they always had access to the people and information most important to them. Similar levels of agreement were seen in responses to questions asking about participants' being always accessible to others and information. In other words, there was general agreement that participants were constantly connected to their social, work, and informational networks. Thinking in terms of presence, this means participants perceive the potential for being interrupted at any given moment of their day. On one hand, this connectivity seemed to generate increased perceptions of productivity, but respondents also indicated difficulty focusing their attention and being reminded of stressful obligations, which could lead to negative effects on task performance and mental health. The findings are clearer about smartphones being one of the most regular focuses of respondents' attention. Large majorities reported checking for notifications or alerts at least five or more times per hour (or every 12 minutes). There seems a strong likelihood that this level of engagement will have significant consequences for individuals' ability to focus their attention, maintain awareness of their physical surroundings and interact with others and information in their physical environments.

Participants in the 2020 survey largely mirrored the presence sentiments of those in the 2015 sample except for their responses to a question asking about the "do not disturb" mode on their smartphones. The 2020 respondents showed a 30% increase in the use of this functionality over the 2015 participants. The small sample size in 2020 weakens the ability to place too much emphasis on this difference, but it suggests the population began to engage in increased presence management strategies in the years between the surveys. Indeed, a similar proportion (57%) of 2020 respondents indicated that they regularly turn off or ignore their smartphone for at least an hour each day, which supports the view that some individuals do actively work to manage how much attention they allocate to their smartphone each day.

Despite the supporting evidence for individuals working to manage the access and accessibility generated by smartphones, most participants associated smartphones with increased levels of distraction and frustration. Most participants felt like their smartphone hindered their ability to focus and found it difficult to stop using their smartphone when other tasks required their time or attention. Participants were also frustrated by others focusing on their smartphones when they were trying to have conversations. That is, despite some participants' efforts to manage smartphone distractions, most respondents indicated their ability to control their attention and the attention of others suffered because of smartphone use.

When asked about the effect of smartphones on specific aspects of participants' identities, perceptions varied. While most participants felt smartphones enhanced their ability to engage with others and information, almost half of the respondents did not feel like smartphones improved their performance as students or family members. Larger majorities felt like smartphones improved their performance as citizens and friends. Future studies with larger samples would need to confirm these results, but these findings seem to indicate functions that require deeper levels of engagement and higher quality of attention like scholastic achievement and intimate relationships suffer more from smartphone use than the maintenance of weaker social ties and general information gathering behaviors.

When asked about the effects of smartphones on their productivity in general, participants expressed varied, and at times conflicting, points of view. There was widespread agreement that smartphones enhanced the ability to multitask and opened opportunities to work when and where respondents desired. However, large majorities reported they would be both more productive by staying off their smartphones and less productive if they did not have a smartphone. This finding reinforces the need for future studies to explore the tasks and roles for which smartphones are perceived as beneficial tools and for which users perceive them to be hindering distractions.

The presence findings are far clearer about smartphones occupying a pervasive and fundamental role in respondents' lives. Respondents almost universally start their day by checking in with their device and keep their smartphone nearby while sleeping. Most said it was common to check for smartphone notifications in the middle of the night and that they experienced anxiety when separated from their smartphone. These results strongly support the view that smartphones have permeated virtually every aspect of participants lives. Assuming these trends continue, understanding how individuals define themselves and their surroundings will require an understanding of how identity and information is consumed through these devices.

Table 3

Presence Questions and Response Frequencies in 2015 and 2020

Survey Question	2015	2020
I feel like my smartphone makes it "a lot" easier to stay in touch with the people I care about	99%	n/a
I feel like my smartphone makes it "a lot" easier to be productive while doing things like standing in line	78%	83%
I feel like my smartphone makes it "a lot" harder to give people my undivided attention	78%	74%
I feel like my smartphone reminds me of stressful obligations when I am away from work or school	62%	66%
I feel like my smartphone makes it "a lot" harder to focus on a single task that requires my full attention	75%	83%
I regularly use the "do not disturb" mode on my smartphone	22%	55%
On average, I check my smartphone for notifications or alerts at least five or more times per hour	80%	71%
I feel like I can always be contacted by my friends	93%	n/a
I feel like I can always be contacted by members of my family	97%	n/a
I feel like I can always be contacted by my employer	90%	n/a
I feel like I always have access to the information I need to do my schoolwork	98%	n/a
Regardless of where I am, I always have access to the people I care most about	94%	n/a
Regardless of where I am, I always have access to the information I care most about	89%	n/a

Table 3 Continued

Survey Question	2015	2020
My smartphone is the most important tool I have for communicating with others	n/a	89%
I get frustrated when others use their smartphone while we are having a conversation	n/a	77%
I often check my smartphone for notifications even when I did not notice an alert (e.g., vibration or sound)	n/a	86%
It is common for me to check my smartphone for notifications when I wake up in the middle of the night	n/a	54%
I keep my smartphone near my bed when I go to sleep at night	n/a	94%
I check my smartphone for notifications immediately after waking up in the morning	n/a	89%
My smartphone makes it harder to focus in class	n/a	71%
My smartphone makes it easier to multitask	n/a	77%
I get anxious when I do not have or forget my smartphone	n/a	66%
During my waking hours, I regularly turn off or ignore my smartphone for at least one hour each day	n/a	57%
Life would be less stressful without smartphones	n/a	49%
Overall, my smartphone helps me be a better friend	n/a	74%
Overall, my smartphone helps me be a better student	n/a	54%
Overall, my smartphone helps me be a better citizen	n/a	60%

Table 3 Continued

Survey Question	2015	2020
Overall, my smartphone helps me be a better family member	n/a	54%
I often get overwhelmed by the obligations I have each day	n/a	83%
It is difficult for me to give my attention to everyone who requests it each day	n/a	66%
My smartphone makes it a lot easier to work from home	n/a	74%
My smartphone gives me the ability to work when it is convenient for me rather than during predetermined times of the day	n/a	69%
I would be less productive if I did not have a smartphone	n/a	43%
It is difficult to keep up with all the information I receive each day	n/a	80%
It is difficult for me to stop using my smartphone even when I know I should be working on other tasks	n/a	80%
I often keep using my smartphone at night even when I know I should go to sleep	n/a	89%
I would be more productive if I could stay off my smartphone for a couple of hours each day	n/a	83%
I would make fewer silly mistakes if I did not check my smartphone so frequently	n/a	60%

Note. Percentages represent proportion of respondents that agreed with each survey question.

Descriptive Statistics Related to Publicity Perceptions and Behaviors

This section presents findings about how smartphone use related to perceptions of publicity. Generally, questions in this section sought to understand if smartphone use lead individuals to perceive themselves as having more public (i.e., less private) identities. The names of some social media platforms changed between the 2015 and 2020 surveys (due to changes in popularity and/or existence). There were two questions in the 2015 survey dropped after analysis deemed responses difficult to interpret. These questions were revised and included in 11 questions added to the 2020 survey that were not asked in 2015. This section begins will a discussion of the 2015 results and notes similarities among the 2015 and 2020 respondents. It concludes with a discussion of the responses to the questions added to the 2020 instrument. A full listing of publicity question response frequencies is shown in Table 3 below.

Overall, participants reported high levels of activity on social media platforms. This seemed to contradict to the low levels of confidence expressed about social media platforms being secure places to share personal information. Nonetheless, at least two-thirds of respondents reported posting to three or more social media sites monthly in both the 2015 and 2020 samples. There was also general agreement that respondents perceived others searching for personal information about them online. Respondents in both surveys were less likely to report engaging in similar search behaviors themselves, but this was likely a result of social desirability bias (Turner & Martin, 1985). Interestingly, respondents were almost evenly spilt on willingness to exchange their personal information for discounts or services. That is, notable numbers of respondents in both samples said they were unwilling to (knowingly) allow the use of their personal information in exchange for the connectivity and social capital made possible by social networks. It is worth noting here that 92% of respondents in 2015 and 100% of respondents in 2020 posted to at least one social media account each month, In other words, despite nearly half of the respondents saying they were unwilling to trade personal information for free services, virtually all the respondents also reported engaging in precisely this behavior.

Respondents were asked directly about the performance of identity online and in physical settings by the 2015 survey. Most of the participants felt like they performed a "public identity" in both settings, but the researcher was surprised to find more respondents said they performed a public identity in physical settings (where they presumably had less ability to edit and filter their presentations) than in online settings. As a result, these direct probes were replaced by several more nuanced questions in the 2020 survey. Using these revised measurements, strong majorities of respondents said both they and others presented the "best" versions of themselves online. Conversely, slightly less than half of the 2020 respondents agreed that it was easier to be yourself online than it was in physical settings like class, restaurants, bars and/or church. Again, the small 2020 sample must be noted, but these results support a view that the presentation of self in digital environments tends to be idealized and carefully tailored. If true, it also means individuals are consuming idealized presentations of others' identities on social media which may skew expectations and desires toward unrealistic standards.

Perhaps the most salient finding from the questions added to the 2020 survey was the prominence of smartphones as the medium by which social media sites are accessed. Virtually all the respondents said their smartphone was the most common medium used for accessing their digital social networks and a similar majority said they would be less frequent contributors to social media if they did not own a smartphone. This finding represents another tension in the responses as participants seemed to be simultaneously aware of the risks posed by both social media platforms as well as smartphone sensor technology, yet they also reported frequently using these platforms on a regular basis and in combination with one another. The present study can only speculate about causality, but it provides circumstantial evidence that smartphone and social media use have become so pervasive that individuals are simply unable or unwilling to reflect their stated desires for privacy with their behaviors.

Table 4

Publicity Questions and Response Frequencies in 2015 and 2020

Survey Question	2015	2020
I upload or share information to an Instagram account at least one time each month	78%	74%
I upload or share information to a Twitter account at least one time each month	66%	57%
I upload or share information to a Facebook account at least one time each month	63%	20%
I upload or share information to a YouTube account at least one time each month	22%	37%
I upload or share information to a Tumblr account at least one time each month	22%	n/a
I upload or share information to a Pinterest account at least one time each month	21%	6%
I upload or share information to a LinkedIn account at least one time each month	10%	9%
I upload or share information to a Google+ account at least one time each month	8%	n/a
I upload or share information to a Vine account at least one time each month	6%	n/a
I upload or share information to a Snapchat account at least one time each month	n/a	86%
I upload or share information to a TikTok account at least one time each month	n/a	31%
I upload or share information to a Reddit account at least one time each month	n/a	23%
I upload or share information to a Facebook Messenger account at least one time each month	n/a	17%

Table 4 Continued

Survey Question	2015	2020
After I meet someone new, I usually search for them online	58%	63%
I assume people search for information about me online	75%	69%
I am willing to share some personal information about myself with websites or apps in order to get free or discounted services (e.g. coupons, free shipping, and free accounts or services)	53%	54%
I have a public identity that I perform online	61%	n/a
I have a public identity that I perform in public places like stores, restaurants and bars	69%	n/a
I often get stressed out by what people post on social media	n/a	46%
Society would be less divided without social media	n/a	69%
It is easier to be yourself online than it is in physical spaces like class, restaurants, bars and/or church	n/a	49%
Sometimes, I express viewpoints I do not agree with online to see how others will react	n/a	14%
I have separate social media profiles for close friends on which I share information I would not want member of my family or acquaintances to see	n/a	63%
I would post to social media less frequently if I did not have a smartphone	n/a	91%
I would be less informed about what is happening in my community if I did not have a smartphone	n/a	86%

Table 4 Continued

Survey Question	2015	2020
I get a lot of satisfaction form the responses I receive to my social media posts (e.g., likes)	n/a	57%
I present the "best" version of myself on social media	n/a	69%
Most other people only post the good parts of their life to social media	n/a	86%
My smartphone is the most common way I access social media platforms	n/a	94%

Note. Percentages represent proportion of respondents that agreed with each survey question.

This chapter has highlighted how smartphones have become the primary media device in the lives of the study's participants. A large majority of participants reported using their smartphone more than any other medium. This is an important finding as smartphones were a minority medium only eleven years ago (Pew Research Center, 2019).

The findings above have also shown mixed perceptions about privacy in relation to smartphone use. Most participants expressed at least some level of concern about the security of their personal information, but a notable percentage did not share this concern. Furthermore, most participants reported behaviors like sharing location data and regularly using social media which seem to be conscious decisions to share their personal information with platforms most participants view as being insecure. Overall, participants said they were confused about how their information was being used by smartphone applications and most had given up on trying to understand the information sharing policies of big data industry entities. Instead, participants seemed to have focused their attention on controlling which individuals in their social networks had the ability to view their content rather than trying to manage how industry or government agencies may use their information.

The effect of smartphones on presence perceptions was largely observed through the lens of access and accessibility to others and information. Participants reported both enhanced productivity and functionality by having "always on" access to others and information, but the potential to be interrupted by their smartphones was also perceived as being disruptive. As with privacy, participants seemed to be somewhat confused about how they felt about the impact of smartphones on their presence and/or focus. Respondents simultaneously reported the perception that they would be less productive without smartphones and more productive if they would do a better job managing how much time they spent using their smartphones.

Paradox was a theme in the publicity findings as well. Despite low confidence in social media information security and an unwillingness to exchange personal information for goods and services, essentially all respondents were frequent contributions to social media platforms. Participants also reported using their smartphones as the primary medium for accessing social media, which implies participants choosing (either consciously or unconsciously) to make large volumes of their personal information available to platform developers and their third-party partners. The following chapter will build on these findings by investigating statistical relationships between factor variables and speculating about potential causes for the observed relationships.

Chapter 6: Inferential Statistics and Modeling Results

This chapter will present results from the 2015 and 2020 surveys using inferential statistics. Primarily, correlations between factor variables will be presented. Results from significance testing using the 2015 data are given first and used as a baseline for comparison with findings in the 2020 data. As the additional questions in 2020 yielded more than double the number of factors found in 2015, factors judged to be similar in both samples are presented first and factors unique to the 2020 data are presented last.

Multiple regression models used to test the hypotheses articulated in the methods section conclude the chapter. In addition to models created to test the impact of privacy concern, attention, and public identity as predictors of smartphone use, several secondary models were also created to reflect new questions and curiosities that arose between the 2015 and 2020 data collections. Specifically, models that made smartphone use and gender predictors of factors like privacy concern were generated to look for potential effects of smartphone use rather than at how different attributes and perceptions may influence smartphone behaviors. Readers may find the tables of factor compositions in Appendices I through L useful throughout this chapter.

2015 Factor Variable Correlations

Correlations between the nine factor variables, gender, and the total amount of smartphone use reported by the 2015 respondents were run and analyzed using SPSS. The risk of type I error is relatively low with 11 variables. Specifically, it is expected that $\left(\frac{N+N-1}{2}\right) \times \alpha$ significant correlations would be found due to Type I errors or

$\left(\frac{11+10}{2}\right) \times .05 \approx .5$. In other words, it would be expected to see up to one significant correlation due to repeated measures of the 2015 variables. Reporting for all measured correlations is found in Table 4 below.

Overall, most of the observed factor relationships were intuitive. Respondents reporting higher levels of confidence about the security of their smartphone applications were generally less likely to report higher levels of privacy concern. Similarly, respondents who expressed higher levels of concern about the safety of the personal information were more likely to take actions to protect their personal information from being accessed and used by applications on their smartphones. Highlighting the relationship of privacy and publicity was the observation that respondents who engaged in higher levels of privacy protection behaviors were less likely to engage in publicity behaviors like posting to social media and thinking others search for information about them online. In other words, the 2015 results supported the idea that higher levels privacy concern and protection behaviors were associated with lower levels of publicity perceptions and behaviors.

Presence findings were also intuitive. As one would expect, participants who perceived higher levels of access to others and information also perceived themselves as being more accessible to others and information. Furthermore, respondents indicating higher levels of smartphone distraction also reported higher levels of negative effects from smartphone notifications. This was additional support for an understanding of smartphones as creating costs and benefits for user's attention. Smartphones are perceived to be aides to productivity and functionality while also making users vulnerable to increased interruption by others and/or notifications on their smartphones.

Table 5

2015 Factor Variable Correlations

Factor Name	1	2	3	4	5	6	7	8	9	10	11
1. Secure Platforms	—	-.293[**]	-.175	-.066	.084	-.133	.031	.048	.166	.039	-.026
2. Privacy Concern	-.293[**]	—	.284[**]	-.059	-.174	.095	.089	-.029	-.028	-.031	-.106
3. Protection Behaviors	-.175	.284[**]	—	-.133	-.181	-.083	.051	-.212[*]	-.147	-.183	-.107
4. Accessibility to Others	-.066	-.059	-.133	—	.623[**]	.047	-.081	.151	.046	.012	-.056
5. Access to Others	.084	-.174	-.181	.623[**]	—	-.037	-.141	.243[*]	.189	-.083	-.069
6. Attention/Distraction	-.133	.095	-.083	.047	-.037	—	.288[**]	.133	.030	-.006	.209
7. Notification	.031	.089	.051	-.081	-.141	.288[**]	—	-.052	.175	.068	.023
8. Publicity	.048	-.029	-.212[*]	.151	.243[*]	.133	-.052	—	-.058	.291[**]	.157
9. Performance	.166	-.028	-.147	.046	.189	.030	.175	-.058	—	-.128	.025
10. Gender	.039	-.031	-.183	.012	-.083	-.006	.068	.291[**]	-.128	—	.271[*]
11. Sum Smartphone Use	-.026	-.106	-.107	-.056	-.069	.209	.023	.157	.025	.271[*]	—

Note. Correlations are between factor variables with outliers on smartphone use level removed ($n = 86$). Gender was coded zero for males and one for females.
[*]$p < .05$. [**]$p < .01$.

2020 Variable Correlations

Correlations were computed between the 20 variable structures identified in 2020, gender and total level of smartphone use. With the increased number of correlation calculations, came a slightly higher risk of type I error. Specifically, it was expected that $\left(\frac{22+21}{2}\right) \times .05 = 1.075$ erroneous significant correlations were identified due to type I error. That said, the increased detail made possible by the additional questions and resulting factors generated an opportunity to observe more nuanced relationships between respondents' perceptions of privacy, presence and publicity. Generally, it was observed that participants' perceptions of privacy control and information security were often related to their perceptions of smartphone utility and levels of smartphone distraction. Furthermore, higher levels of smartphone use and those reporting habitual use smartphone use behaviors were associated with higher levels of public identity beliefs and behaviors. These relationships are discussed below with a full table of 2020 factor correlations shown in Table 5.

A theme observed in the 2020 correlations was a relationship between privacy perceptions and the perceived negative consequences of smartphones. At a top level, those that perceived application security or who felt competent in managing their application functions, tended to view fewer negative smartphone effects in terms of distraction and identity management. Higher levels of perceived application security were associated with higher levels of public identity behaviors and lower concern about censoring or being selective about what was posted to social media profiles. Those that felt competent about managing their smartphone privacy settings were also less likely to report being interrupted by their smartphones, feeling like they were overloaded with information, being disconnected from others, and needing to be selective about what they shared on social media. The findings only allow for speculation, but users who are more capable of controlling the flow of information on their smartphones may experience (or at least perceive to experience) fewer negative consequences because of their use.

On a similar note, those who reported higher levels of social media concern or being more selective about what they shared on social media were also less likely to express feeling like their smartphones generated distractions from "always on" connectivity. Not surprisingly, they were also less likely to report behaviors associated with public identity, which in this study were highly tied to participants' levels of posting to social media. That is, those who reported what might be termed implicit privacy protection behaviors were less likely to perceive some negative effects of smartphone use. This seemed to be important as participants who perceived smartphones as enhancing their productivity or functionality also reported lower levels of lost focus and feeling disconnected from others because of smartphone interruptions.

On the other hand, those that perceived smartphones as having negative impacts on their privacy or who had seemingly given up on trying to control how their personal information was collected and used by smartphone application developers were more likely to perceive negative effects from smartphone use. Those reporting higher levels of privacy apathy were less likely to report perceptions of smartphones being tools that improved their ability to function or be productive. These respondents were also more likely to perceive smartphones as being responsible for disconnecting them from others when they were interacting in face-to-face contexts. Mirroring these perceptions were those that felt like they had lost control of their personal information because of smartphones. These users were also more likely to report smartphones hurting their ability to focus, decreasing their productivity, and disconnecting them from others. Again, it is speculation, but this may help explain the observation that those who more strongly desired additional regulations to limit how their personal information is collected and used by the big data industry and governments were also more likely to report high levels of privacy apathy, lost focus due to smartphone use, habitual smartphone use, and obligation stress.

A final note about the 2020 findings relates to perceptions of publicity or public identity and negative consequences of smartphone use. Those that scored higher on the public figure or public identity variable were more likely to report habitual smartphone use behaviors and negative consequences from constant smartphone connectivity. Habitual smartphone use behaviors were also associated with higher levels of personal identity being derived from social media profile maintenance. Perhaps most notably, the public figure variable was the only factor to correlate with total smartphone use, with higher levels of public identity being associated with higher use levels. The small 2020 sample size and the inability of the present findings to make causality assessments needs to be kept in mind, but all of these findings seem to indicate a relationship between smartphone use levels and identity creation through social media platforms. Specifically, the results indicate that smartphone use may be associated with higher levels of public identity development which could be continually reinforced and maintained by the

habitual engagement made possible by the always on connectivity of smartphones.

Table 6

2020 Factor Variable Correlations

Factor Name	1	2	3	4	5	6	7	8	9	10	11	12	13	14	15	16	17	18	19	20	21	22
1. Security	—	-.150	.023	.048	-.187	-.451**	-.436**	-.179	.080	.073	.102	-.175	-.270	.164	.017	.101	.461**	.058	-.010	-.058	.312	.233
2. Regulation	-.150	—	.495**	-.132	.327	.067	.231	-.235	.463**	.374*	.215	-.064	.339*	-.212	.359*	.186	.002	.167	.313	-.032	.184	.000
3. Apathy	.023	.495**	—	-.325	.286	.022	-.049	-.402*	.486**	.280	.273	-.036	.150	-.097	.444**	.253	.115	.162	-.052	-.141	.180	.172
4. Competence	.084	-.132	-.325	—	-.259	.207	.143	.183	-.243	-.236	-.429*	-.387*	-.222	.183	-.490**	-.006	-.244	-.021	-.369*	-.047	-.482**	-.165
5. Lost Privacy	-.187	.327	.286	-.259	—	.097	.180	-.109	.408*	.029	.514**	.265	.151	-.418*	.437**	-.094	.023	.315	.163	-.074	.094	-.101
6. Social Concern	-.451**	.067	.022	.207	.097	—	.243	.168	-.191	-.105	-.379*	-.127	-.002	.025	.030	-.143	-.372*	.141	-.219	-.103	-.290	-.175
7. Privacy Worry	-.436**	.231	-.049	.143	.180	.243	—	-.276	.044	-.079	-.176	.082	-.033	-.109	-.053	-.218	-.211	.194	-.234	-.062	-.393*	-.143
8. Functionality	-.179	-.235	-.402*	.183	-.109	.168	-.276	—	-.477**	-.111	-.229	-.114	-.242	.114	-.425*	.313	-.084	-.154	.115	-.179	-.047	-.137
9. Lost Focus	.080	.463**	.486**	-.243	.408*	-.191	.044	-.477**	—	.535**	.431*	.223	.324	-.063	.465**	.010	.460**	-.011	.335	-.038	.289	.277
10. Habitual	.073	.374*	.280	-.236	.029	-.105	-.079	-.111	.535**	—	.208	.239	.207	.137	.226	.345*	.525**	.029	.351*	.031	.441**	.153
11. Connection	.102	.215	.273	-.429*	.514**	-.379*	-.176	-.229	.431*	.208	—	.339*	.289	-.197	.421*	-.020	.350*	.087	.415*	-.093	.412*	.214
12. Overload	-.175	-.064	-.036	-.387*	.265	-.127	.082	-.114	.223	.239	.339*	—	.229	-.096	.093	-.088	.184	-.160	.027	-.006	.263	.113
13. Obligation	-.270	.339*	.150	-.222	.151	-.002	-.033	-.242	.324	.207	.289	.229	—	-.185	.237	.163	.171	.242	.392*	.260	.187	.195
14. Productivity	.164	-.212	-.097	.183	-.418*	.025	-.109	.114	-.063	.137	-.197	-.096	-.185	—	-.226	.403*	.211	-.076	-.059	-.280	-.026	.188
15. Disconnected	.017	.359*	.444**	-.490**	.437**	.030	-.053	-.425*	.465**	.226	.421*	.093	.237	-.226	—	-.258	.190	.162	.322	-.107	.385*	-.045
16. SM Identity	.101	.186	.253	-.006	-.094	-.143	-.218	.313	.010	.345*	-.020	-.088	.163	.403*	-.258	—	.263	.212	.185	-.021	.232	.112
17. Public Figure	.461**	.002	.115	-.244	.023	-.372*	-.211	-.084	.460**	.525**	.350*	.184	.171	.211	.190	.263	—	.112	.352*	-.093	.510**	.507**
18. Social Stress	.058	.167	.162	-.021	.315	.141	.194	-.154	-.011	.029	.087	-.160	.243	-.076	.162	.212	.112	—	.138	.067	-.168	-.140
19. Presentation	-.010	.313	-.052	-.369*	.163	-.219	-.234	.115	.335	.351*	.415*	.027	.392	-.059	.322	.185	.352*	.138	—	.058	.401*	.088
20. Authentic	-.058	-.032	-.141	-.047	-.074	-.103	-.062	-.179	-.038	.031	-.093	-.006	.260	-.280	-.107	-.021	-.093	.067	.058	—	.089	-.108

Table 6 Continued

Factor Name	1	2	3	4	5	6	7	8	9	10	11	12	13	14	15	16	17	18	19	20	21	22
21. Gender	.312	.184	.180	-.482**	.094	-.290	-.393*	-.047	.289	.441**	.412	.263	.189	-.026	.385*	.232	.510**	-.168	.401*	.089	—	.176
22. Smartphone	.233	.000	.172	-.165	-.101	-.175	-.143	-.137	.277	.153	.214	.113	.195	.188	-.045	.112	.507**	-.140	.088	-.108	.176	—

Note. Correlations are between factor variables with outliers on smartphone use level removed (*n* = 34). Gender was coded zero for males and one for females. Factor names were abbreviated for formatting: Security = Perceived Security. Regulation = Desire for Regulation. Apathy = Privacy Apathy. Competence = Privacy Competence. Lost Privacy = Lost Control of Privacy. Social Concern = Social Sharing Concern. Functionality = Enhanced Functionality. Habitual = Habitual Behavior. Connection = Constant Connection. Overload = Information Overload. Obligation = Obligation Stress. Productivity = Organized Productivity. Social Stress = Social Media Stress. Presentation = Selective Social-Presentation. Authentic = Authentic Sharing. Smartphone = Sum Smartphone Use Level.
*p < .05. **p < .01.

Multiple Regression Model Results

The following section presents multiple regression models that were used to test the hypotheses articulated in the methods chapter. Specifically, the models tested what (if any) ability reported perceptions about privacy, presence and publicity had for predicting the level of time individuals spent using their smartphones. To model contested and somewhat vague concepts like privacy, presence and publicity, factor variables were selected that were believed to operationalize these concepts in terms with which most readers would agree. Results of the 2015 model are presented first before discussing results of the 2020 model. Since questions were altered between the two surveys, some of the longitudinal reliability is lost, however the researcher believed the underlying concepts being measured were consistent in both models and were assumed to be valid "snapshots" of beliefs about privacy, presence, and publicity as well as total levels of smartphone use in both samples.

In the 2015 model, the privacy concern factor was selected as the variable to represent privacy perceptions that would be likely to impact smartphone use. Given the power and sensitivity of smartphone sensor technologies to generate large volumes of highly detailed user information, it was hypothesized that higher levels of privacy concern would be associated with lower levels of smartphone use (i.e., have a negative beta in multiple regression modeling). The attention/distraction factor was selected as the variable to represent presence perceptions that may impact smartphone use. Since smartphones provide "always on" access and accessibility to others and information, it was hypothesized that higher levels of perceived distraction from smartphones would be predictive of higher levels of smartphone use (i.e., have a positive beta). Finally, the opportunistically named publicity factor was selected as the variable to represent publicity perceptions in the 2015 model. Those who engaged in more public forms of communication were thought to be more likely to embody perceptions of themselves as having a public identity. Since smartphones make it easy to publicly disseminate information about oneself via sensor technologies and ready access to social media platforms, it was hypothesized that higher levels of publicity perceptions would be predictive of higher levels of smartphone use (i.e., have a positive beta). The results of the 2015 model are summarized in Table 6 below. Appendix I provides a table of all the 2015 factor variables.

The model shows results of standard multiple regression modeling where all three predictors along with a gender control variable were entered simultaneously. The overall 2015 model was statistically significant, however only one of the hypothesized relationships was supported. Although all the predictor variables had beta values consistent with predictions, only the presence factor (attention/distraction) was statistically significant. Specifically, respondents' perceived distraction from smartphones was a significant predictor of smartphone use, with higher levels of perceived distraction being predictive of more time spent using smartphones (and vice versa). Gender was also a significant predictor. Being female significantly predicted higher amounts of time using smartphones and being male predicted a lower total amount of time spent using smartphones. That is, H1 and H3 were not supported by the 2015 results. Only H2 was supported, and although significant, the full model explained a relatively small amount of the total variance in reported time spent using smartphones. Only 13% of the total variance was explained by the predictor variables.

To examine the unique predictive capability of presence perceptions and gender, t ratios for each predictor's slope (b) was evaluated. The unique variance explained by distraction/attention perceptions and gender were obtained by squaring their part correlations. The $sr^2 = .04$ for attention/distraction perceptions and $sr^2 = .06$ for gender. In other words, gender was the strongest predictor of smartphone use in the model, but neither of the variables were strong predictors of total smartphone use.

Although not statistically significant, the direction of the relationship between privacy concern and publicity behaviors with smartphone use level was consistent with predictions. The negative slope of privacy concern indicates higher levels of concern being associated with lower smartphone use levels. The positive slope on publicity behaviors indicates higher smartphone use by those engaging in activities that support a public identity. However, as these are insignificant indicators, they could simply be the result of chance. Overall, the model explained a relatively low level of the total variance in smartphone use but pointed to attention/distraction perceptions being a topic relevant for future study about how smartphones may be impacting individuals' daily lives.

Table 7

Predictors of Total Smartphone Use in 2015

Variable	B	95% CI
Constant	137.637	[-.133.475, 408.749]
Privacy Concern	-11.386	[-31.553, 8.781]
Attention/Distraction	26.489[*]	[.713, 52.265]
Publicity	3.881	[-12.744, 20.506]
Gender	91.531[*]	[14, 169.062]
R^2	.134	
F	3.134[*]	

Note. $N = 86$. CI = confidence interval. Constant term is undefined as zero was not a possible score for variables in the model. Adapted from the Publication Manual of the American Psychological Association, 2010.
[*]$p < .05$

2020 Multiple Regression Model Predicting Smartphone Use from Factor Variables

The multiple regression model of the 2020 data attempted to mirror the 2015 model as closely as possible given the changes in phrasing and additional questions included in the 2020 survey. Again, factor variables were used to operationalize the predictor variables measursing privacy, presence and publicity. Privacy worry was selected as the factor that most closely resembled the privacy concern factor in 2015. Thus, it was hypothesized that higher levels of privacy worry would be associated with lower levels of smartphone use (negative beta). The lost focus factor was evaluated to be the closest representation of the attention/distraction factor used in 2015. As such, it was hypothesized that increased perceptions of lost focus would be associated with higher levels of smartphone use (positive beta). Finally, the public figure factor was thought to be a close approximation of the publicity factor used in the 2015 model. It was hypothesized that public figure perceptions would be positively related to smartphone use levels. Thus, the 2020 factors and gender were used to predict the total amount of smartphone use reported by the 2020 participants. Results of the full model are presented in Table 7 below. The 2020 factor variables are shown in Appendices J-L.

Mirroring the 2015 model, the 2020 results used standard regression analysis with all four predictor variables entered simultaneously. Using this method, the overall model was significant, but showed a notable shift in variable performance. Given the small sample used in 2020, the model results need to be interpreted with caution. That said, presence perceptions were no longer significant predictors of smartphone use in the 2020 results. Instead, the only significant predictor of smartphone uses in 2020 was publicity or the public figure factor. Consistent with the hypothesized relationship, higher levels of publicity perceptions were associated with higher amounts of time using one's smartphone. That is, H3 was supported by the 2020 results, however H1 and H2 were not. Also of note was the increased predictive capability of the full model, which explained 28% of the variance in the time participants reported using their smartphones. This was over twice the amount of variance explained by the 2015 model.

In order to understand the unique predictive capability of the public figure factor, a t ratio of its slope was computed in order to generate a squared measure of its part correlation ($t(33) = 2.647$, $p = .013$, $sr^2 = .17$). The public figure factor was shown to uniquely explain about 17% of the total variance in smartphone use, which was more than the full model explained in 2015. The remaining predictors were not significant, and their relationship to smartphone use levels could be the result of chance. However, their slopes were consistent with the 2015 results. Privacy worry had a negative slope and the lost focus predictor had a positive slope. Overall, the 2020 results need to be confirmed with a larger sample, but suggest a noticeable shift in smartphone effects. Evidence of a relationship between publicity perceptions grew tremendeously between the 2015 and 2020 data collections. On the other hand, the relationship between access and accessibility distractions and smartphone use seemed to disappear over the same time. The following chapter will discuss and speculate about this shift in greater detail, but a final finding about the public figure factor and smartphone use is presented as a follow up analysis below.

161

Table 8

Predictors of Total Smartphone Use in 2020

Variable	B	95% CI
Constant	24.41	[-511, 560]
Privacy Worry	-11.17	[-52, 30]
Lost Focus	4.36	[-15, 24]
Public Figure	36.56*	[8, 65]
Gender	-68.81	[-247, 110]
R^2	.277	
F	2.779*	

Note. $N = 34$. CI = confidence interval. Constant term is undefined as zero was not a possible score for variables in the model. Adapted from the Publication Manual of the American Psychological Association, 2010.
*$p < .05$

Revised 2020 Multiple Regression Modeling

The technological capability and use of smartphone had grown rapidly between the time of the two data collections in this study (Pew Research Center, 2019). With these increases in mind, the author began to reconsider the directionality of the relationships being examined between smartphone use levels and variables measuring privacy, presence, and publicity. Given the growing ubiquity of smartphone use, it seemed less and less valid to believe privacy/publicity beliefs or perceived levels of smartphone distraction would prevent or encourage individuals to use a smartphone. Instead, it seemed likely that smartphone use had become a commonplace element of modern life.

Consequently, the author flipped the direction of the questions posed in 2015 to investigate how smartphone use may be impacting beliefs and concerns about the privacy of personal information, perceived levels of distraction and public identity perceptions. That is, the author wanted to understand what predictive capability smartphone use levels may have on the variables initially modeled to predict smartphone use.

The result of the revised questioning was the creation of multiple regression models that used smartphone use levels to predict privacy, presence and publicity perceptions while controlling for gender. Using this formulation, the only model in which smartphone use was a significant predictor was for public figure perceptions and behaviors. Results of this model are shown in Table 8 below. Smartphone use explained approximately 18% of the total variance in public figure perceptions and behaviors, $sr^2 = .179$, $t(33) = 3.154$, $p = .004$. Gender was also a significant predictor of the public figure variable explaining another 18% of the variance in responses, $sr^2 = .183$, $t(33) = 3.184$, $p = .003$. The strength of this model is notable. Together, the variance in total smartphone use level and gender explained 44% of the total variance in public figure perceptions which are essentially perceptions and behaviors that are counter to privacy or protecting one's personal information. Again, the small sample and the bounded nature of the public figure factor limit the reliability and validity of these findings. That said, in combination with the 2020 model above, the 2020 data strongly indicate a relationship between publicity perceptions and smartphone use that was not present in the 2015 findings.

Table 9

Predictors of Public Figure Variable in 2020

Variable	B	95% CI
Constant	8.871**	[6.82, 10.93]
Sum Smartphone Use	.006**	[.002, .010]
Gender	2.774**	[.997, 4.551]
R^2	.44	
F	12.184**	

Note. $N = 34$. CI = confidence interval.
Public Figure variable could range from 3 to 24.
Adapted from the Publication Manual of the American Psychological Association, 2010.
**$p < .01$

The findings in this chapter have highlighted relationships between perceptions about privacy, presence, publicity as well as smartphone use. Many of these relationships were in line with intuitive expectations. Respondents that indicated high levels of privacy concern tended to engage in higher levels of privacy protection behaviors and were less active on social media. Similarly, those that limited their engagement with social media platforms also perceived less distraction from their smartphones. On a deeper level, the findings also showed individuals who reported being apathetic about privacy perceived myriad negative effects from their smartphones including higher levels of distraction, being disconnected from others and feeling overwhelmed by information and alerts. The results supported the view that smartphones increased participants access and accessibility to others and information. However, this increased connectivity could be associated with both increased perceptions of productivity and work flexibility as well as increased distraction and stress. A notable finding was an indication that participants' perceived competence for managing their privacy seemed to be an important intermediary for participants perceiving more positive or negative effects of their smartphone.

Finally, although the findings showed limited support for the study's hypotheses, the shifts observed between the 2015 and 2020 model results point to opportunities for future investigations. The strong relationship between the public figure variable and smartphone use observed in 2020 along with associations between the public figure variable and habitual smartphone use behaviors suggests smartphone use could lead to increased public identity performance. The present results are only speculative on this point and need to be investigated by future studies with more participants, but a potential explanation for distraction becoming less indicative of smartphone is that users have simply come to accept interruptions as part of their daily existence. In this light, perceptions of interruptions from social media and smartphone alerts become a "cost of doing business" for maintaining social capital and identity through smartphones and social media. The following chapter will build on some of the themes developed here and tie the results back to previous studies discussed in chapters two and three.

Chapter 7: Discussion

This chapter will place many of the findings discussed in the preceding two chapters in the context of the theoretical and empirical literatures discussed in chapters two and three. It begins with a discussion of findings about smartphones in general before proceeding to specific findings and concepts related to privacy, presence, and publicity. Generally, the findings in 2020 supported or provided additional insights about the findings in 2015. As such, results are discussed as a whole, but notable differences are noted where applicable. The chapter concludes with a speculative model that combines all the study's findings into a theoretical model of the mobile information cycle. The model is intended to give myriad points of entry for future studies to expand and improve upon this study's findings.

At the most general level, the rapid rise in smartphone use and the need to study the effects of smartphone use in the population at large are clearly supported by the findings. The average time participants spent using their smartphones grew on average by more than two hours in the five years between data collections. Participants were already heavy smartphone users in 2015, but the nearly eight hours per day spent using smartphones by the 2020 respondents is a clear indicator of how quickly this new medium has become mainstream. It is important to remember than barely one-third of adults owned a smartphone as recently as 2009 (Pew Research Center, 2019). The rapid growth in adoption and rise in use levels alone would make the smartphone a worthy medium of study for all modern media researchers, but when combined with the rapid evolution of smartphone technology as both a consumption and information production device, the smartphone is poised to rewrite understandings of communication and mass media theory. A particularly salient finding was the rise not only in the time participants spent using smartphones, but the doubling of the time they spent using smartphones to access and contribute to social media platforms. The combination of smartphone sensor capabilities and incentives for big data entities to capitalize user information make this trend particularly important for understandings identity management and consumption behaviors in the age of smartphones (Friedman, 2005; Hanslip, 2020; Kietzmann et al., 2018; Makridakis, 2017; President's Council of Advisors on Science and Technology, 2014).

The study's results about privacy and smartphone use support and provide nuance to many previous studies and concepts. Primarily, the results support both Mulligan et al.'s (2016) understanding of privacy as a contested term as well as Nissenbaum's (2011) theorizing of contextual privacy. The limited questionnaire used in 2015 identified four different privacy variables and the questions added in 2020 brought this number to nine. Thus, an important contribution here is a variety of lenses through which privacy may be investigated in relation to smartphone use. Although privacy concern and/or worry was the primary focus in this study, equally valid questions may investigate how different applications augment sharing behaviors, particularly encrypted applications such as Signal and WhatsApp. Alternatively, future studies may investigate how perceived levels of privacy competence relate to observed sharing of personal information with application developers.

Most respondents in both samples expressed concern about how their personal information was being collected and used by organizations. This concern was seen in several different ways. Participants were selective about which applications to install on their devices and about how they shared information on social media. Most directly, participants expressed privacy concern when they explicitly asked about their level of worry concerning the security of their personal information. That said, respondents also spent large amounts of time using their smartphones to access social media despite expressing very low confidence in these platform's ability to protect their information. Furthermore, most 2020 respondents felt their attempts to control the use of their personal information had minimal effect. Large majorities took it for granted that their personal information was tracked by governments and businesses. To a certain extent, these findings are consistent with observations of the privacy paradox (Norberg et al., 2007). However, it is also important to keep in mind the relative lack of controls users have for protecting their privacy. Tools like those developed by Acquisti et al. (2017b), Liccardi et al. (2014) and Schaub et al. (2015b) that help users interpret and select apps that protect their privacy have not been adopted by either of the major application marketplaces. Furthermore, tools like the personal privacy assistant developed by B. Liu et al. (2016b) render many popular applications unusable when required privacy permissions are revoked. In this light, the lack of control expressed by respondents in the present study and in previous studies is understandable if not unavoidable (Auxier et al., 2019).

In the absence of tools to control how their information is shared with businesses and governments, users seem to turn to tools that are available for controlling which members of their social networks (or known audience) have access to their information. Respondents readily engaged in screening and/or restricting who could see their social media postings, however findings by Brandimarte et al. (2013) suggest that these screening behaviors likely result in even risker information disclosures. It is believed that users discount how much information they are sharing with big data entities due to enhanced perceptions of control over which individuals in their networks have access to their posts. That is, by giving users control over which nodes in a network have access to an individuals' information, users in turn forgot about monitoring by the network itself and the resulting inhibition leads to greater amounts of overall information sharing. This reasoning may be useful for understanding the perceived paradox between participants' high levels of privacy concern and behaviors that would seem to jeopardize their privacy. In fact, participants seem to acknowledge this discrepancy as most of the respondents indicated a desire for additional regulations to protect their personal information.

Whereas perceived privacy made some participants more comfortable, those who desired more regulations to protect how smartphones could use their personal information reported a wide range of undesirable effects. Those who more strongly desired additional privacy regulations were also more likely to report high levels of smartphone anxiety, being overwhelmed by smartphone notifications and feeling like family relationships suffered because of smartphone use. These respondents seemed to be experiencing tension between the connectivity and attendant interruptions generated by smartphones and the desire to focus their attention on elements in their physical environments such as close friends and family members. Thus, this study's findings support a somewhat more nuanced understanding of the autonomy paradox theorized by M. Mazmanian et al. (2013). Respondents seemed to be negotiating between perceptions of smartphone connectivity enhancing convenience and connectivity and degrading focus and physical intimacy. However, respondents who at least perceived more control over their smartphone's privacy settings were less likely to report many of the negative consequences to autonomy and focus. That is, it seemed that perceptions of effective privacy management techniques mitigated many of the negative effects on respondents' ability to be present to others and desired tasks. Yet, given the high levels of smartphone use and particularly the use of smartphones for accessing social media, the present study suggests a more nuanced interpretation. By enacting the relatively limited number of privacy control mechanisms on social media (e.g., which members of a network can access certain user content), participants are actually giving social networks even more personal information in the form of social preferences and intimate information they are only comfortable sharing with select individuals (while forgetting about network surveillance). When viewed from this lens, it seems likely that all respondents are frequently being interrupted and distracted by smartphone notifications, but those that give the networks more detailed information then receive more targeted and seemingly more relevant notifications/interruptions. As Fischer et al. (2010) showed, the perceived usefulness and relevance of interruptions has significant impacts on whether users interpret negative or positive effects from smartphone notifications. While somewhat speculative in the present study, this interpretation helps to alleviate tensions not only between privacy concerns and behaviors, but also perceived distraction levels and privacy competence perceptions.

In fact, the results suggest the big data industry may benefit from users perceiving control over their personal information. The present findings showed that once a respondent became apathetic about privacy, they were much more likely to interpret negative smartphone effects such as decreased productivity, lower focus, and being disconnected from others. Given these perceptions, it was not surprising to see privacy apathy also strongly associated with a desire for more regulation to protect privacy as a means for helping alleviate these concerns and effects. To maintain access to the wealth of user information made possible by smartphone data collections, the big data industry needs to walk a narrow line of giving users enough control to perceive security while not giving them so much control as to limit the amount of information available to industry entities and partners.

Tensions and conflicts were also seen in participants' perceptions of smartphone utility. Most participants expressed conviction that their smartphone made them more productive and better multitaskers. Participants believed the information convenience and interpersonal connectivity made possible by smartphones enhanced their roles as citizens and friends. However, participants were more divided about the effect of smartphones on their roles as family members and students. Prior researcher has shown that merely engaging in multitasking behaviors leads to lower task performance (Carrier et al., 2015; Jeong & Hwang, 2016). Indeed, even this study's participants overwhelming agreed that they would be more productive if they could decrease the time they spent using their smartphone and that they had trouble disengaging from their smartphones when they had other tasks that needed their attention. A takeaway here is the always on connectivity of smartphones has fundamentally changed how participants negotiate presence in their daily lives. They have access to others and information at virtually every waking moment and can theoretically transform any environment into social or work settings as they choose. At the same time, the connectivity allowed by smartphones creates opportunities for outside entities to always interrupt users and redirect their attention to social or commercial purposes regardless of their location in physical space. Respondents in 2020 seemed to be taking actions to control this access and accessibility by using "do not disturb" tools and outright ignoring their smartphones for periods of the day, however when compared to their 2015 counterparts, they also showed a clear trend toward more daily use of their smartphones, not less.

As smartphone use increased between the 2015 and 2020 surveys, so too did reported use of social media. The average time spent using smartphone for social media doubled between 2015 and 2020. In both samples, large majorities contributed to three or more social media profiles each month, and in 2020, 100% of respondents posted to at least one account monthly. Yet, in both samples, almost half the respondents indicated they were unwilling to share personal information in exchange for free goods or services. Furthermore, strong majorities indicated a belief that social media platforms were not secure platforms for sharing personal information. Again, the results point to a paradox between expressed beliefs and reported behaviors. In this case, it seemed respondents could be reconciling this tension by focusing on the carefully tailoring the information they shared on social media rather than their concern about how their smartphone and

social media applications themselves could be collecting, analyzing, and commodifying

their personal information. The results also suggest that this negotiation is influenced by

the feedback received from likes, shares and/or comments to their sharing behavior. In

sum, it seems likely that individuals are both performing and consuming identity on

social media and through their performance and consumption behaviors they are

providing marketers and application developers with unprecedented insights about their

tastes and attitudes (Kietzmann et al., 2018). While the 2020 sample makes strong

generalizations impossible, the trend indicated is a rapid growth in smartphone use for

social media engagement, which if true, could fundamentally alter individuals

perceptions of themselves and their role in society (Meyrowitz, 1985).

At first glance, the regression modeling of the 2015 and 2020 data present two

different understandings about how smartphone use may be affecting users. Although the

2015 modeling explained relatively low levels of variance in smartphone use, and

smartphone use did not appear to predict any significant variance in privacy, presence or

publicity perceptions, there was a significant relationship between the distraction variable

and smartphone use. The fact that distraction levels stood out as significant predictors of

smartphone use levels in 2015 but not in 2020 raised the question of what led to this

change. For comparison, the overall proportions of responses about smartphones causing

distractions or impairing focus were roughly eight percentage points higher in the 2020

data, but despite this rise, the lost focus variable was not significantly associated with

smartphone use levels. Given the much smaller sample in 2020, this could be the result of decreased statistical power, but it is also possible that distractions from smartphone have become so ubiquitous that there is no longer enough variance to associate it with variance in smartphone use levels. Instead, the 2020 data shows a clear and much stronger relationship between public identity perceptions and smartphone use levels. Indeed, this relationship was observed using multiple measures. Significant statistical relationships were observed between the correlation between public identity and smartphone use levels, with public identity as a predictor of smartphone use in the presence of gender, privacy concern and lost focus scores as well as when smartphone use was used as a predictor of public identity perceptions in the presence of gender. Another important change observed by comparing the 2015 and 2020 results was the doubling of time spent using smartphones on social media applications between data collections. Social media use was far and away the most popular smartphone use in 2020 jumping up from the third most common activity in 2015.

Explaining the Contradictions

To put these results in context, it is important to remember the potential for smartphones to infer personal information and preferences about their users using sensor technologies found on most smartphones in combination with algorithmic filtering and machine/artificial learning technologies (Makridakis, 2017). The present findings make clear that respondents are increasingly engaged in a cycle of information sharing and consumption that connects understandings of the big data industry with the attention economy. A speculative model of how this cycle functions is shown in Figure 1 and explained below.

Figure 1

Smartphone Information Echo Chamber

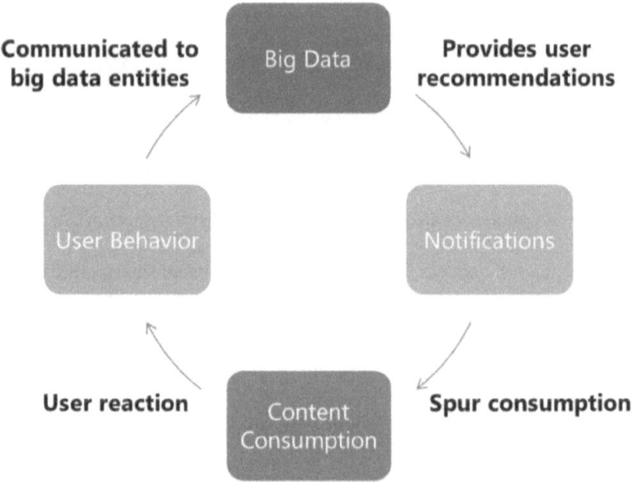

In an attempt to make sense of many of the contradictions and tensions evidenced by this study's results, the cycle shown in Figure 1 is suggested as a framework for how smartphone users may become unconscious participants in a smartphone information echo chamber (Bruns, 2017; Cinelli et al., 2021). The cycle begins with virtually any activity conducted on a smartphone (user behavior), but the actions conducted on social media are particularly relevant. All of the social media sites identified by the respondents, and especially the top three most popular sites (Snapchat, Instagram and Twitter) are for profit platforms that generate the vast majority of their revenue through the sale of targeted advertising. Smartphone actions are logged by application developers and used by big data analysts to generate unique user profiles by which advertising messages can be selectively disseminated. These messages are typically communicated through notifications or through integration into users' information feeds/scrolls/interfaces. Users' reactions to these notifications (clicks, time spent viewing, likes/dislikes, shares, etc.) generate insights that further refine users' profiles which increase the value of the information developers can sell to advertisers (i.e., the ability to charge for more message "targets"). Once begun, this cycle repeats without end, constantly updating and refining itself to tailor notifications and content to users' preferences which ideally stimulate more consumption and more insights about users' consumption behaviors.

Thinking about this cycle and the models generated by the 2015 and 2020 survey results helps to explain the observed shifts in the explanatory value of privacy, presence, and publicity variables. As a contribution to understandings about the ways smartphones are impacting the lives of users, the figure above shows smartphones have transformed every user into a role that resembles a broadcaster. That said, the smartphone broadcaster differs from the traditional notion of the term in several important aspects. Primarily, whereas radio and television broadcasters were very conscious of their audience and went to great lengths to understand their consumers' demographics and behaviors (i.e., audience research and ratings), smartphone users are largely unaware of how the information they broadcast into social media and other smartphone applications is consumed, analyzed, and used by the big data industry. In both the 2015 and 2020 samples, vast majorities reported high levels of concern about the security of their personal information, but the additional 2020 probes highlight how users overwhelming acknowledge that they have little control over how businesses and governments use this information (and in many cases have given up on privacy entirely). Instead, it appears users focus their attention on the elements of sharing that they do feel control over like which members of their intended audiences (i.e., friends and followers) get to see their contributions while largely ignoring the much larger "audience" of advertisers and developers behind the scenes. In other words, smartphone users appear to be tailoring their messages to specific individuals in their networks, but their unintentional sharing of their contacts/networks themselves, location and travel data, purchase histories, content consumption behaviors/reactions, browsing histories, job histories, education levels and purchasing power represent the information most coveted by their "audiences." Thus, in a strange reversal of the mass communication cycle, it is the "feedback" from audiences that is highly targeted and "narrowcast" back to users in the form of recommendations, notifications and advertisements.

Thinking about smartphones as transforming users into individual broadcasters is a useful lens for understanding several of the study's other findings as well. For starters, it is an intuitive way to understand why the public figure variable was so prominent in the 2020 results. As social media use through smartphones skyrockets, individuals are correct in thinking that others are actively searching for information about them and there is a very high demand for them to contribute their thoughts, actions, and options to social networks. These are the very thoughts and feelings being encouraged by the industry trying to capture as much of their attention as possible. Obviously, most users are concentrated on receiving feedback from "friends" in their networks, but each input is nonetheless communicated to a vast array of big data players that use the information to encourage more content consumption and contributions on their platforms (which includes connecting them with "like" others who are more likely to give them positive and likeminded feedback).

The disappearance of the relationship between smartphone uses and distractions levels may also be the result of the increasing sophistication of social networks to provide users with content that is compelling and valued by users (Fischer et al., 2010). Indeed, the overall level of lost focus effects went up between 2015 and 2020, but as was discussed, as information becomes more relevant to users, the less likely they are to avoid notifications, regardless of if doing so distracts from completing other tasks or even interferes with healthy living behaviors. In other words, the personalized information made possible by smartphone sensors, network modeling and demographic profiles generates feedback that user simply can mot or do not want to ignore.

In such an environment, it is much easier to understand why privacy concern and even perceived competence bears little relationship to smartphone use levels. While one may be highly concerned about how their information is being used by application and platform providers, when information is being tailored so specifically to user's tastes, wants, and needs, the incentive to seek out alternative information that may run counter to the users' personal tastes and preferences is substantially decreased if not nullified. That is, despite one's concern, they are constantly being reminded that their concerns are shared by many others but many people just like them are contributing just as frequently to the same platforms (Cinelli et al., 2021). Simply put, the comfort, ease, and convenience (i.e., perceived benefits) of social media and smartphone applications outweigh any vague concerns about privacy violations which occur largely outside of users' view and their ability to understand.

This paints a somewhat different portrait of the effects of increased digital connectivity theorized by McLuhan (1964) and Meyrowitz (1985). Rather than a global village in which the central nervous system of each individual is thrust into contact with others of different cultures, beliefs and resources, it seems smartphones are more likely to create a seemingly endless sea of cultural centers with no concept of others' or viewpoints. They are also rendering notions of front and backstage behaviors obsolete in an objective sense. Although individuals are clearly still performing the "best" versions of themselves to their intended audiences on social media platforms, the full performance of all their private information is being observed, consumed, and used to produce highly targeted goods and services as users devote more and more time to social media applications on their smartphones. That is, users place in society is changing, but not to reflect more informed access to information formerly hidden from them, but as performers of identity that is reinforced and measured for the purposes of marketing and consumption.

Clearly, there are advantages and disadvantages to the information landscape be created by smartphones and other increasingly pervasive IoT technologies. There are many advantages to understanding the needs of individuals and tailoring solutions for those needs. The caution being highlighted here is that individuals seem to have lost control over how to regulate the personal information being given to parties beyond their intended audiences and there are strong economic incentives for platforms to use this information for maximizing organizational advertising revenues rather than solving individual and societal problems. It is one thing to broadcast published and produced content across the public airwaves for any interested party to consume. It is quite another to broadcast every detail of one's daily life to parties motivated by profiting off its sale and utilization.

What are the Solutions?

One of the key findings from both the 2015 and 2020 respondents is the disconnect expressed between privacy concern and privacy protections offered to smartphone users. A central argument being made is that the lack of privacy protections has resulted in smartphone users being highly vulnerable to manipulation by big data players. The current legal paradigm governed by notice and consent is simply inadequate. The presented findings make clear that users want to limit the exposure and use of their personal information, but the notice and consent model makes it all but impossible to keep up with an ever-changing landscape of technological capability and changes in the legal language used when updating application software. As a result, it was not surprising to find that many users have simply given up on keeping track of how their smartphone applications are using their personal information. While the solution to this issue is multifaceted, reasonable proposals include shifting responsibility to corporations to respect users' desires once they have been expressed (e.g., a one-time do not track request that applies in perpetuity between users and businesses), removing the ability for privacy disagreements to be relegated to arbitration rather than legal damages, the inclusion of privacy protection officers in top levels of corporate hierarchies, and giving users legal ownership of their personal data and information (Fairfield, 2011; Janeček, 2018; Schwartz, 2016). How to balance and implement aspects of these recommendations is in urgent need of careful evaluation and debate, but what has been made clear is the current framework is fundamentally flawed and failing to support users wishes and interests. Indeed, social media business models are antithetical to protecting users' privacy. As smartphone technology continues to evolve, the potential for invasion and manipulation of users' information will continue to grow.

A related benefit of regulating the big data industry to respect the privacy desires of users would likely be a reduction in the amount of polarization and division currently being fueled by the profiling made possible by smartphones and digital social networks (Gillani et al., 2018). If one were to limit the tailoring of information networks to everyone's minute preferences, there is a real potential for an overall improvement in the social fabric of communities and individuals' mental health. Nevertheless, many experts believe the potential for this future is quickly disappearing, and if action is not taken soon, could be gone forever (Pew Research Center, 2020). No matter what, there is strong evidence in the present study and from the views of these experts that the need for media literacy, or the ability to think not only about the sources and credibility of information but also about the technological role in the selection and visibility of information, will be of paramount importance as digital surveillance technologies continue to grow in sophistication.

The benefits of new privacy regulations combined with a focus on fostering media literacy extend beyond the realm of privacy and information control to attention management strategies as well. As individuals become more aware of the attention economy and its incentive to capture as much user attention as possible, there is reason to believe individuals will take action to regain some control over the frequency and constancy with which they interact with their smartphones and other IoT technologies. Again, managing attention in the face of nearly constant demand for it will not be a one size fits all model, but will likely be a combination of user awareness/literacy, notification management tools, and implementing protections for individuals engaging in problematic or excessive smartphone use behaviors (Mehrotra, Hendley, et al., 2016; Panova & Carbonell, 2018; Park et al., 2017). What cannot happen is the continued reliance on the patently flawed notice and consent framework to inform users about how their information is being collected and used. The data make clear that this will simply lead to greater availability of private information which seems poised to create higher levels of distraction, polarization, privacy apathy and confusion.

Study Limitations

The findings and arguments above should be understood within the limitations of the study and in context of the data collected. Using survey instruments that largely collected participants' perceptions represents hypothetical or cognitive reasoning rather than objective behavioral measurements and practices. As with any survey, responses could reflect idealized or socially expected levels of concern and behavioral characteristics. These effects were combated by ensuring participant anonymity but can never be completely overcome. As previously mentioned, Adjerid, Peer et al.'s (2016) findings have shown that hypothetical privacy concern levels in surveys may overestimate actual concern at the point of data collection, and the present study is no exception.

The relatively small *n* size of the 2015 sample, and certainly the 2020 sample, limit the power of model interpretations. It is likely larger samples would generate significance for additional variables related to privacy, presence, and publicity variables. This is far from certain though, and the present study is best thought of as two snapshots of media students in 2015 and 2020 that point to trends that will need to be monitored and examined in the larger population to validate and refine. The findings and arguments concerning public identity perceptions and related concerns about privacy protections are particularly relevant for future research and validation. The proposed smartphone echo chamber points to numerous future lines of inquiry as well.

The selection of media students is less a limitation of the study than a limitation on generalizing the findings to the population at large. While it is reasonable to assume the responses reflect good estimations of undergraduate perceptions about smartphone use relationships with privacy, presence and publicity factors, the relationships within the population at large and especially older and less educated demographics could be (and likely are) very different. The present study was interested in understanding perceptions and behaviors of young and savvy media users, and consequently cannot speak to populations less fluent with media technologies and digital social networking. However, less savvy users are likely to have lower levels of media literacy which would likely mean a population with increased vulnerability to manipulation by big data practices.

In sum, the study was designed as an examination of how well hypotheses generated from earlier media effects theorizing by McLuhan (1964) and Meyrowitz (1985) held up to today's media communication environments. The results point to both works maintaining relevance, but needing revised understandings added to compensate for the vast amount of user data and analytic analyses that are now commonplace in mediated communication. Privacy, publicity, performance, and attention factors are still very relevant areas of study for media effects researchers, the challenge will be ascertaining these effects in an era when even savvy media consumers will be challenged to truly understand how their information and behaviors are interacting with computers to develop beliefs about security, identity and society.

Future Research

As with most studies, the findings presented generate more questions than they answer. Although the principal component analyses (PCAs) in Appendixes C-H were primarily conducted to uncover latent variables for the regression modeling, their results also provide direction for future studies. Privacy concern or the level of individuals' worry/desire for protection of their personal information has been the focus of much of the digital privacy research to date (this study included). Understanding users' concern levels is important, but it is also clear that users have different thoughts about the level of privacy protections offered across different categories of applications (i.e., security perceptions). It is also clear that privacy can be addressed from the vantage point of which actions users choose to take to limit exposure of their private information. The desire for anonymity is another aspect of privacy in digital environments. There is strong potential for deeper study of anonymity through the lens of examining contexts in which users desire anonymity, how anonymity may alter users' disclosures and how anonymity may impact or enhance the viability of social business models and how anonymity may alter individuals' exposure to information counter to their previously held beliefs.

The delineating of specific information sharing behaviors that users may be inclined to allow or prohibit are also missing from the present study. Future work may seek to understand if users are truly differentiating between the "release" of information to others in their network and the ability for third parties to "access" the information they supply during their digital communications. Understanding the limits of user comfortability with various data profiling practices would be beneficial for both lawmakers and businesses themselves as communities attempt to balance economic development with individual needs for autonomy and privacy. Indeed, understanding the extent to which users may be inclined to engage with the ownership of their personal data will be essential to drafting the needed regulations for using private information.

The dimensionality of presence questions also generates potential research questions. Smartphones have been observed to have dramatic impacts on individuals' abilities to access others and information as well as their availability or readiness to be contacted by others and information. The results of the study suggest the capabilities afforded by smartphones could be interacting with privacy and public identity perceptions, but there are other interactions which warrant future investigation including how smartphones alter relationship development, the needed adaptation of smartphones to facilitate their use as learning tools and how gratifications from the increased capacity for access and accessibility may be influencing behavior and self-perception.

Perhaps most pointedly, the relationship between smartphone use and attention/distraction levels merits much more investigation. The results lead to a belief that attention or the ability to focus is decreasing as social media and smartphone use increases. Assuming this is true, understanding which tasks and/or relationships are most affected will be important moving forward. It also points to the conflict of interest between users who presumably wish to be able to control how their attention is allocation and application designers who desire to attract as much consumer attention as possible. As availability increasingly becomes part of the service package provided by organizations and industries, we need to understand if delivering availably may ultimately undermine our ability to solve complex problems and form deep relationships with others. Furthermore, investigation is needed at the practical and not just cognitive levels. Further study of the relationship between smartphone notifications and task performance, propensities to engage in multitasking behaviors, stress/mental health and/or memory/comprehension levels should guide the future development of tools and strategies for maximizing cognitive capability while unlocking the positive potentials of smartphones as tools for generating insights about societal and individual level problem solving.

The skyrocketing of social media use observed in the five-year period (i.e., public identity perception and performance) seems very likely to be having a significant impact on self-perception and how members of digital networks perceive each other. Combined with understandings of how all digital and physical/digital hybrid behavior is monitored, analyzed, and sold by smartphone applications, the true effects of smartphone use on public and private identity development is almost certainly profound, but beyond the scope of the present study. Some resent work (Barry et al., 2019; Henderson, 2018) has begun to speak to these issues, but as smartphones and embedded computing environments continue to grow in sophistication and prevalence, these issues will remain critical for the foreseeable future.

Finally, the findings suggest there could be a difference in how users perceive the ability to be "authentic" in online versus offline environments. Future work may ask if digital environments really do generate identity behaviors that run counter to expectations and/or perceived limitations in physical world environments. No matter what, the effectively infinite amount of information being delivered by smartphones is rapidly altering understandings of privacy, personality, and cognitive capacity. As computing and its associated surveillance capabilities become more and more pervasive, the development of individuals' sense of self and their communities will continually evolve alongside them. The pace of these changes has been rapid and will only accelerate moving forward. It is imperative that we work quickly to understand the effects of a new communication medium that in little more than a decade has become the dominant communication platform for the next generation and has the potential to rewrite the rules of social engagement and individual identity.

Chapter 8: Conclusion

This work began with an introduction to the theory of media effects and an argument for differentiating between the effects of media content and the devices through which content is transferred. As has been discussed, the interplay between these two concepts is substantial, but their differentiation is important. It has been argued that the smartphone, a medium that has only come into prominence over the past decade, has the potential to fundamentally alter the way individuals think about themselves, their environments, and their communities. Given the recent rise in both smartphone adoption and utilization patterns, an investigation into the effects of this new medium was justified. Specifically, it was argued that the multifaceted sensor technologies contained on most modern smartphones, combined with the tendency for individuals to always carry their smartphones on their person, created unprecedented opportunities for data collection and utilization when compared to traditional mass media like the radio, television and even desktop computing.

After arguing the need for smartphone research, the investigation was grounded in media effects theories created by Marshall McLuhan (1964), Walter Ong (1975), Joshua Meyrowitz (1985), and Dana Cuff (2003). Extending these theories' predictions and observations to the smartphone era lead to an investigation of how millennials were managing privacy, presence, and publicity in the face of digital surveillance and "always on" smartphone connectivity. The researcher tested hypotheses about how privacy, presence and publicity understandings were evolving to reflect the connectivity and information sharing made possible by smartphones. It predicted privacy concern would hinder the relative amount of time individuals spent using their smartphones (H1). It also predicted that the connectivity of smartphones would make it more difficult for individuals to focus their attention (H2). Finally, it predicted that increased levels of smartphone use could be predicted by increased perceptions of individuals trying to maintain a public identity or high information sharing activities (H3).

Previous empirical studies related to media effects theorizing were presented in the literature review. Through a review of the literature, it was established that all three hypotheses included concepts that were contested or had elements of vagueness in their definitions. The potential for smartphones to collect user data was explored in detail and attention was given to the costs and benefits made possible by using smartphones as data collection devices. Evidence was presented that strongly suggested smartphone users were often unaware of, or misinformed about, how their smartphone collected information about them. Furthermore, studies of informed users showed that even experts often disagreed about the meaning of many smartphone information policies which called into question the notice and consent legal framework that currently governs the collection and use of smartphone data.

In addition to privacy concerns, results of studies examining the effects of interruptions and multitasking on task performance and relationship management were reviewed. It was observed that individuals generally overestimate their ability to multitask and that interruptions had negative effects on quality and efficiency of task performance. This was relevant to the present study because the increased connectivity made possible by smartphones increased the potential for individuals to be interrupted by and/or to distract themselves with the content readily available through their smartphones.

To investigate the potential effects of smartphones, a survey design was presented and tested using focus groups and a pilot study. After incorporating feedback, two surveys were administered to undergraduate students majoring in media arts and studies. The first survey was administered in 2015 and the second in 2020. Samples for both data collections were drawn from students in the same media class at the same university. However, due to the rapid evolution of smartphone technology, and empirical studies published between the two collections, several items were revised and added to the survey instrument used in 2020. Although doing so hurt some of the longitudinal validity of the study's findings, their inclusion also brought significant gains in the detail and clarity of respondents' perceptions. Based on previous empirical survey research and methodological guidance, both samples met minimum qualifications for being viewed as representative of media undergraduates' beliefs and behaviors related to smartphone use, privacy, presence, and publicity.

Since many of the study's concepts involved contested or multifaceted terms (i.e., privacy, presence, and publicity), modeling and inferential statistics of the results relied on factor analyses that identified different dimensions of each concept being investigated. To test the study's central research question and hypotheses, factor variables were identified that measured privacy concern/worry, attention/distraction and/or lost focus, and publicity and/or public identity. Narrowing these concepts to factor definitions strengthened the reliability of the variables and provided additional insights about the ways participants thought about the central concepts in the study.

The results overwhelmingly supported an understanding of smartphones being central to the lives of participants. Not only have smartphones become the primary media vehicle for content consumption, but they also appear to be the primary means by which individuals connect with others in their personal and professional networks. In other words, smartphones have blended many former distinct media into a singular and central medium for information consumption and production. There are myriad consequences to this synthesis, but the present study has highlighted a few key insights about impacts to privacy, connectivity, and public identity. Primarily, participants are conflicted between their desire for privacy and to engage with platforms that exist to harvest their personal information. Large majorities of participants reported being concerned about the security of their personal information on social media platforms, yet by 2020 the average participant spent more than two hours using their smartphone for social media activities. It may be the case that participants were expressing greater levels of concern than they actually held due to response bias (K. Martin & Shilton, 2016), however it may also be the case that social pressure to participate on social media platforms negates or discounts levels of concerns. Regardless, participants do seem to experience the privacy paradox (Norberg et al., 2007) even in a population that theoretically has a deep understanding of media technologies (i.e., undergraduate students in media). This leads to serious questions about media literacy and a potential for extending its definition to include media devices as well as content. Much like media effects theories, discussions of media literacy often center around identifying "fake news" and developing critical skills for analyzing media *content* (Bulger & Davison, 2018; Mason et al., 2018). What has hopefully become clear is that technologies themselves are also central to understanding and making informed decisions about the new media environment. Being able to identify propaganda or patrician content may be important, but equally important is the realization that media users are often the products being sold by social media platforms. The finding that undergraduates immersed in the study of media are so readily discounting the collection and sale of their personal information is alarming. With enough personal information, these platforms gain the ability to subtly nudge and provoke users toward content and beliefs that may come from legitimate sources, but nonetheless reinforce limiting points of view that are ultimately designed to serve the interests of platform advertisers and partners far more than to enhance users mental, physical and economic wellbeing. The call here is to include more content and discussion in the educational process (particularly of media students) around the collection, use and sale of individuals personal information which continues to grow in volume and sophistication due to the proliferation of smartphones and other IoT technologies.

In a similar fashion, respondents seem torn between perceiving smartphones as enhancing their ability to be productive at more times and in more places and feeling like smartphones make it harder to focus on tasks that require their attention or to disengage from content consumption activities when they want to be focused on completing other tasks. The suggestion here is that respondents are simply trying to resolve their cognitive dissonance between smartphone use and productivity. We know that humans are terrible multitaskers (Adler & Benbunan-Fich, 2012; Carrier et al., 2015; Jeong & Hwang, 2016), but large majorities of respondents are checking their phones every 12 minutes. Respondents themselves seem to paradoxically acknowledge this fact as more than 80% believed they would be more productive by staying off their smartphone, despite 57% believing their smartphone made them more productive. It was also astounding to see 90% of respondents delaying sleep due to smartphone use and 80% saying they continued using their smartphone even when other tasks were demanding their attention. Collectively, this findings support understandings that smartphone use can be associated with addictive behaviors (Y.-H. Lin et al., 2015), regardless of whether these behaviors meet more clinical definitions of the term (Panova & Carbonell, 2018).

Smartphones and social media are facilitating the development of a global village in the sense that individuals have seemingly infinite access and accessibility to others and information. However, it challenges the notion of this developing a global community. Instead, it has argued that smartphones have turned every individual into a revolutionary sort of broadcaster. Rather than broadcasting to the public, the smartphone users' largest audience is a network of information technologies that is consuming information about users well beyond what they intend to communicate. These information networks generate carefully tailored feedback loops designed to spur users toward additional disclosures/broadcasts that can be used in a never-ending cycle of consumption and user profiling. That is, the ramifications of digital connectivity are in some ways radically different than what early theorists like Marshall McLuhan had predicted. The most notable finding of this study showed that extending ourselves into cyberspace with mobile devices carries with it the potential to create an infinite number of social/societal centers that prioritize and reinforce narrow understandings of one's self and their surroundings. That is, instead of a singular global village, we seem to be moving toward a world of a million tribes, each perceiving themselves to be the "majority" opinion as social contacts and information are tailored to mirror one's tastes and preferences. Although the convenience of smartphones offers many benefits, there are clear indications that users are discounting costs to their attention and autonomy.

Respondents overwhelming indicated they are going to use their smartphones regardless of their levels of privacy concern. To date, companies have taken advantage of this fact by collecting vast amounts of users' data using legal agreements based on notice and consent frameworks. Users are generally unaware of how public their behaviors on smartphones are to third parties or perceive this access differently than the "public" at large being able to access their information. Nonetheless, smartphone users have essentially become public figures to many application developers and these companies are in many cases freely using users' data as a standalone product and as a means to better align users' interests with their own products.

In total, the need for two related tools was identified. First, users' privacy concern levels showed the need for something akin to copyright protections for individuals' private information. Just as copyrighted creative works can only be used in specific fair use contexts or as the result of licensing agreements, so too should user information default to limited use by companies. Users are unable to enact their privacy desires in pervasive surveillance environments and as the sophistication and ubiquity of sensor technologies continues to evolve, users will continue to have their agency degraded. It would be dangerous and irresponsible to continue asking users to navigate the world of big data. The need for new privacy regulations combined with a focus on generating increased media literacy among the public at large cannot be overstated.

Secondly, there is a need for smartphone applications to work in conjunction or through a centralized service to manage and deliver notifications to users in ways that maximize user productivity and minimize interruption. Obviously, users themselves desire the connectivity and responsiveness of smartphones, but there is a demonstrated need for better user understanding of smartphone effects and tools to help them manage an infinite universe of content and connections. Regrettably, there is not a specific remedy suggested here, but rather a call for experiments on how notification managers, privacy protections, and/or assisted downtimes may affect attention, task performance and stress levels in users. It may be that a single user opt-out/in can be used to govern all websites and apps associated with a specific user or that individuals are given some kind of dashboard to manage what (if any) personal information they are willing share with third parties. Central to any proposal, however, is the need for users to maintain application functionality regardless of their desires to share personal information. This may (and likely will) result in users having to "pay" for many services that are currently perceived as being free. Yet, this is exactly the point, perhaps by forcing users to confront the value of their personal information and the relatively small level of benefits given in return to them for this information, many of the current issues stemming from social media polarization and information manipulation will begin to be addressed by regulations and new business models.

That said, the results represent user perceptions and self-reported behaviors. In this regard, future studies need to evaluate performance levels of users across different levels of observed smartphone interactions and with different levels of observed digital interactions/behavior to better understand how the perceptions reported here align with real world behaviors and outcomes. Smartphones have become ubiquitous and habitually used by sampled participants. Understanding how these tools are affecting individuals from a psychological and societal lens will be essential to understanding how identity, health and productivity are being shaped in the modern media environment increasingly dominated by smartphones and IoT technologies. The results reported here are hoped to be prompts and suggestions for future inquiry of a society still at the beginning of a frontier that will be shaped and permeated by media technologies capable of transforming individuals' understandings of themselves and the environments they occupy every day.

www.ingramcontent.com/pod-product-compliance
Lightning Source LLC
LaVergne TN
LVHW042105190726
843493LV00006B/1369